MODULAR CHEMISTRY

Revision Aid for Module CH4

P G Blake & K D Warren
Formerly Chief Examiners

AS/A LEVEL

WJEC AS/A Level Modular Chemistry
Revision Aid for Module CH4

Published by the Welsh Joint Education Committee
245 Western Avenue, Cardiff CF5 2YX

First published 2001

Printed by Hackman Printers Ltd
Clydach Vale, Tonypandy, Rhondda, CF40 2XX

ISBN: 1 86085 472 9

Revision Aids for WJEC A/AS Level Modular Chemistry

As a result of the Government initiated overhaul of the A/AS Level Examination system, a substantial revision of the existing syllabus (now to be called a specification) was undertaken by the WJEC during 1999. Much of this revision related to rules and module weightings but some alterations in content and content sequence were also needed and the authors have therefore been asked to update their guidance booklets (issued under the RND imprint) so that they can be used in conjunction with this new specification (WJEC 2000). This is due to begin to be taught in September 2000 with first examinations in 2001.

The new examination scheme does however differ in a number of important respects from that previously in place. Most significantly the AS qualification is now overtly recognised as constituting a less demanding standard than that corresponding to a full A Level. Consequently the subject core (now to be called the subject criteria) has been divided into two sections, one covering the AS year and the other the A2 year, and is thus designed to deal first with the more fundamental but less complex aspects of the subject, with the more advanced concepts and most (but not all) of the quantitative treatments reserved until later.

As a result of this division the contents of the new modules (CH1, CH2, CH4 and CH5 which correspond roughly to the previous C1, C2, C3 and C4) show a number of alterations from the earlier arrangements. Thus, for example, the treatment of *d*-block chemistry which used to appear in C1, has been transferred to CH5 and replaced by a somewhat fuller treatment of Group II chemistry, most of which formerly appeared in C4. In addition, the coverage of energy changes and equilibrium, together with that of kinetics, has now been split between CH2 and CH5, the more quantitative aspects now being largely reserved for the latter. Finally the subject criteria now require some n.m.r. spectroscopy to feature in the A2 year and this now appears in CH4, along with the other types of spectra previously treated in C3.

Although the revisers have taken advantage of the opportunity to fill a very small number of gaps in the existing syllabus, they have in general followed the principle of minimum perturbation: they have thus changed as little as possible and then only when the very tightly drawn rules and constraints required by QCA (formerly SCAA) made this unavoidable. Much of the content of the four modules CH1, CH2, CH4 and CH5 and the guidance in this revised series of booklets, will therefore be familiar to users of the existing syllabus and cover very similar ground, and the less exacting standard to be applied at AS will, it is anticipated, be brought about largely by a reduction of the level of demand, rather than by the excision of material.

As a result of students being now encouraged to embark initially on (at least) four subjects for AS, it seems likely that rather less contact time will then be available for any given subject. However, the requirement that the overall standard of the full A Level qualification ought to remain unchanged should lead to a rather more intense level of study for the A2 year. This of course is a matter for individual centres each to handle in their own way and, in the main, no attempt has been made in these guides to tailor the subject material to take any account of this. However, throughout CH1 (and later CH2) some use has been made of smaller font size print to indicate sections for which general principles are rather more important than specific detail. The convention of using square brackets to denote non-syllabus extension material, as previously adopted, is moreover also being retained, whilst the meanings of terms such as 'recall', 'describe', 'apply', 'calculate', 'appreciate' and 'show awareness of' are unchanged.

These booklets are thus designed to provide guidance for both students and teachers of the new WJEC A/AS Level Chemistry scheme, but their intention must be reiterated here: they are not designed as text books and still less to replace them and similarly make no claim to give an exhaustive coverage of every aspect of the specification. They are however intended to indicate the salient points to be grasped for each topic and learning outcome and to classify what is expected as regards the depth of treatment needed. As before, the attempt has been made to convey something of the thinking and philosophy guiding the construction of the specification and frequent cross-references are also made to help students appreciate the relationships between various topics and subject areas. For each learning outcome the text indicates what needs to be learned, understood, deduced and applied, and assistance is also given where certain treatments, such as hybridisation, are not required.

In Module CH1 Physical Chemistry necessarily dominates but is supplemented by an introduction to periodicity and examples of main group Inorganic Chemistry. Module CH2 contains further Physical Chemistry, although some of the more numerically demanding material is omitted, and also introduces the basic precepts of Organic Chemistry. The treatment of benzene and aromatic species is now however deferred until Module CH4 and this also treats all the remaining functional group chemistry together with all the stipulated types of spectroscopy. Finally, in Module CH5, all the more detailed treatments of bond strengths, equilibrium and kinetic calculations and redox material are covered, together with further aspects of periodicity and *d*-block Inorganic Chemistry.

The only other change of significance relates to the teaching order of the contents of the A2 year. It has unfortunately proved impossible to retain the situation in which either Module C3 or Module C4 could be taught (and examined) first. It has thus been necessary to adopt the majority preference in which Module CH4 is undertaken first, with Module CH5 to follow thereafter.

Once again the authors hope that this series of Guides may prove of some value to both students and teachers and would continue to appreciate the views of their users. Finally, they would commend to all interested parties the Background Reader produced by the present Chief Examiner, Mr P J Barratt, for the WJEC, which should prove especially valuable to students hoping for high grades or contemplating the new AEA awards, which will shortly supersede the old S Level.

P.G. Blake
K.D. Warren
August 2000

Introduction to Module CH4

With the completion of Module CH2, the material needed to cover the AS syllabus is concluded. The contents of the remaining Modules, CH4 and CH5, are of course based on this basic work in the first part of the scheme, but are independent of each other. In both Modules some new concepts and material are introduced, but at the same time work from the earlier Modules is extended and explored in greater depth.

In this way Module CH4 embraces two main strands: the new concepts relating to spectroscopy are introduced and defined, whilst the organic chemistry, begun in Module CH2, is continued with a detailed coverage of functional group behaviour, of which only a brief awareness was given previously. As regards spectroscopy it is probably no exaggeration to say that no other single area of study – essentially the manner in which atomic and molecular systems interact with electromagnetic radiation – has done more to revolutionise the study of chemistry in the last half century, during which time access to a wide range of spectroscopic techniques has passed from being the preserve of the few, at the frontiers of research, to constituting a facility automatically assumed to be available almost everywhere. In this Module attention is directed to spectroscopy involving both electronic and vibrational excitations and applications to both organic and inorganic systems are included.

As regards the organic chemistry, the functional group coverage is continued with halo compounds, the full range of oxygen containing systems and finally with nitrogen-containing functions, whilst at the end of this Module are considered a range of techniques available for both the synthesis and analysis of organic systems. Included here are the application of infrared spectroscopy to detect the presence of particular bonds or groupings and a general awareness of the use of spectroscopic techniques in analytical problems.

The coverage of organic chemistry in Modules CH2 and CH4 is largely directed towards aliphatic systems, with only a fairly brief treatment of aromatic species. The essential nature of aromaticity is however considered and the coverage of functional group behaviour is sufficient to provide an adequate basis for further, more detailed, study of organic chemistry.

As for Modules CH1 and CH2, non-syllabus background material is shown by its inclusion within square brackets whilst smaller font size print is again used where general principles are more important than specific detail.

MODULE CH4

TOPIC 12 SPECTROSCOPY

- 12.1 The electromagnetic spectrum; frequencies; wavelengths; energies of radiation involved in ultraviolet (u.v.)-visible and in infrared (i.r.) spectra; transitions between available energy levels; emission and absorption spectra.

- 12.2 Atomic spectra - the hydrogen atom spectrum.

- 12.3 Origin of colour; chromophores in organic systems (electronic spectra).

- 12.4 The uses of ultraviolet-visible, infrared and nuclear magnetic resonance spectra in organic chemistry.

Learning Outcomes Topic 12

Candidates should be able to:

(a) recall the energy gradation across the electromagnetic spectrum from u.v. to visible to i.r. spectra;

(b) appreciate the quantisation of energy;

(c) understand the existence of various available energy levels in atomic and molecular systems, restricted to electronic and vibrational levels;

(d) explain the origin of emission and absorption spectra;

(e) appreciate that energy levels can be split by a magnetic field, that certain nuclei, including ^{1}H, possess intrinsic spin, and that measurements of the magnitudes of the interactions between the nuclear spin and the magnetic field are the basis of nuclear magnetic resonance spectroscopy;
(*The required understanding of n.m.r. spectroscopy is limited to the above and that in Topic 17, outcome(d).*)

(f) describe and interpret the visible atomic spectrum of the hydrogen atom (Balmer Series);

(g) recall the direct proportionality between energy and frequency, as implied by $E = hf$, and the inverse relationship between frequency and wavelength;

(No calculations will be set.)

(h) show understanding of the relationship between the frequency of the convergence limit of the Lyman Series and the ionisation energy of the hydrogen atom;

(i) explain why some substances are coloured in terms of the wavelengths of visible light absorbed;

(j) explain the nature of a chromophore and give examples of such groups in organic species, e.g. –N = N– in azo dyes;

(k) use **given** characteristic i.r. vibrational frequencies (expressed in cm^{-1}), to identify simple groupings in organic molecules;

(l) show understanding of the wide applicability of spectroscopic techniques to analytical problems in industry, medicine and the environment.

TOPIC 13　ISOMERISM AND AROMATICITY

13.1　Nomenclature.

13.2　Stereoisomerism - geometrical, optical.

13.3　Benzene: delocalisation energy of benzene; structure and reactivity; electrophilic nuclear substitution; explanation in terms of π electron delocalisation; comparison with alkenes.

Learning Outcomes　　　　　　　　　　　　　　　　　　　　　　　　　　**Topic 13**

Candidates should be able to:

(a)　give the systematic names of all simple compounds, including benzene derivatives, containing the functional groups occurring in this Module;

(b)　understand the term stereoisomerism as embracing both geometrical and optical isomerism;

(c)　explain what is meant by a chiral centre, recall that this gives rise to optical isomerism, and be able to identify chiral centres in given molecules, and understand what is meant by an enantiomer;

(d)　recall that enantiomers rotate plane-polarised light in opposite directions and that equimolar amounts of enantiomers form racemic mixtures;

(e)　calculate the delocalisation or resonance energy of benzene from given enthalpy data;

(f)　describe the structure of, and bonding in, benzene;

(g)　describe and classify the nitration and halogenation reactions of benzene as electrophilic substitution, and recall the mechanism of the mono-nitration of benzene *;

(h)　compare benzene and alkenes with respect to benzene's resistance to addition and explain this resistance in terms of π electron delocalisation.

Note:
* Conditions required.

TOPIC 14 ORGANIC COMPOUNDS CONTAINING HALOGENS

14.1 Formation of halogenoalkanes by direct halogenation of alkanes.

14.2 Nucleophilic substitution of halogenoalkanes by OH^-, CN^- and NH_3. Mechanism of alkaline hydrolysis.

14.3 Comparative alkaline hydrolysis of 1-chlorobutane and chlorobenzene.

14.4 Industrial, commercial and medical uses. Solvents. Refrigerants (CFCs). Anaesthetics. Toxicity. Adverse environmental effects of CFCs.

Learning outcomes   Topic 14

Candidates should be able to:

(a)   describe the formation of a chloroalkane by direct chlorination of alkanes †*;

(b)   (i)   describe the substitution reaction with OH^- and explain this on the basis of the recalled mechanism of the alkaline hydrolysis of 1-bromobutane †*;

    (ii)   describe the substitution reactions with CN^- and NH_3 † *;

(c)   compare the ease of alkaline hydrolysis of chloroalkanes and chlorobenzene and explain the difference in terms of the C – Cl bond strength, and rationalise the greater strength of the C – Cl bond in the latter case;

(d)   show an awareness of the wide use of halogenoalkanes as solvents, the toxicity of some of them, the use of CFCs as refrigerants and in aerosols, and their use in anaesthetics as well as the adverse environmental effects of CFCs;

(e)   understand the adverse environmental effects of CFCs and explain these in terms of the relative bond strengths of the C – H, C – F, and C – Cl bonds involved (c.f. 11(g));

(f)   show an awareness of the use of organohalogen compounds as pesticides and polymers and assess their environmental impact.

Note:
† Balanced chemical equations are required.
* Conditions required.

TOPIC 15   ORGANIC COMPOUNDS CONTAINING OXYGEN

15.1   Alcohols and phenol

15.1.1 Physical properties of alcohols.

15.1.2 Formation of primary and secondary alcohols by hydrolysis and reduction.

15.1.3 Formation of alcohols by addition.

15.1.4 Oxidation, dehydration and other reactions of alcohols.

15.1.5 Ethanol in a social context.

15.1.6 Acidity and reactions of phenol.

Learning outcomes Sub-topic 15.1

Candidates should be able to:

(a) describe the physical properties of the lower alcohols, solubility in water and relatively low volatility, and relate this to the existence of hydrogen bonding;

(b) describe the methods of forming primary and secondary alcohols from halogenoalkanes and carbonyl compounds *;

(c) recall a method for the industrial preparation of ethanol from ethene;

(d) recall:
　(i) the reactions of primary and secondary alcohols with hydrogen bromide, ethanoyl chloride and carboxylic acids (to give sweet smelling esters) †;
　(ii) the dehydration reaction (elimination) of primary alcohols †;

(e) describe the oxidation reactions of primary (7(g)) and secondary alcohols;

(f) show awareness of the importance of ethanol-containing drinks in society, their ethanol content, breathalysers, and the effects of ethanol excess;

(g) show an awareness of the use of ethanol as a fuel;

(h) explain the acidity of phenol and describe its reactions with bromine and with ethanoyl chloride;

(i) recall the colour reaction of some phenols with $FeCl_3$ solution and the use of this test to distinguish phenols from alcohols.

Note:
† Balanced chemical equations are required.
* Conditions required.

15.2 Aldehydes and ketones

- 15.2.1 Formation from alcohols.
- 15.2.2 Relative ease of oxidation of aldehydes. Tollens' reagent. Fehling's reagent.
- 15.2.3 Reduction using $NaBH_4$.
- 15.2.4 Nucleophilic additions.
- 15.2.5 Triiodomethane (Iodoform) reaction.

Learning outcomes   Sub-topic 15.2

Candidates should be able to:

(a) describe the formation of aldehydes and ketones by the oxidation of primary and secondary alcohols respectively (c.f. Topic 15.1);

(b) describe how aldehydes and ketones may be distinguished by their relative ease of oxidation using Tollens' reagent and Fehling's reagent †*;

(c) recall the use of $NaBH_4$ to reduce aldehydes and ketones and the products formed thereby *;

(d) describe the reaction of aldehydes and ketones with 2,4-dinitrophenylhydrazine reagent as a nucleophilic addition-elimination (condensation) reaction and explain the use of this reaction in showing the presence of a carbonyl group and in identifying specific aldehydes and ketones by determining the melting temperatures of the purified products;

(e) describe and understand the mechanism of the addition of HCN to carbonyl compounds as an example of a nucleophilic addition reaction;

(f) describe how the triiodomethane (iodoform) test is carried out and explain its use in detecting CH_3CO – groups or their precursors (c.f. Topic 15.1).

Note:
† Balanced chemical equations are required.
* Conditions required.

15.3 Carboxylic acids and their derivatives

15.3.1 Physical properties and comparative acidity of carboxylic acids.

15.3.2 Formation of carboxylic acids from alcohols and aldehydes.

15.3.3 Conversion to esters and acid chlorides, and the hydrolyses of these compounds.

15.3.4 Reduction of acids using $LiAlH_4$; decarboxylation of acids.

15.3.5 Industrial importance of ethanoic anhydride and polyesters.

Learning outcomes   **Sub-topic 15.3**

Candidates should be able to:

(a)
- (i) describe the physical properties of lower carboxylic acids (volatility and solubility) and relate these to the presence of hydrogen bonding;
- (ii) discuss and show understanding of the relative acidities of carboxylic acids, phenol, alcohols and water, and appreciate that carboxylic acids liberate CO_2 from carbonates and hydrogencarbonates but that phenol does not;
- (iii) recall that phenols in aqueous solution give colour reactions with iron(III) chloride solution;

(b) recall the following listed processes and apply knowledge of them to the elucidation of organic problems:

- (i) the formation of carboxylic acids from alcohols and aldehydes *;
- (ii) the formation of aromatic carboxylic acids by the oxidation of methyl side-chains with alkaline Mn^{VII} and subsequent acidification*;
- (iii) methods of converting the acids to esters and acid chlorides, and the hydrolyses of these compounds †*;
- (iv) the behaviour of acids on reduction with $LiAlH_4$; acid decarboxylation and its use in structure determination *;

(c) recall the industrial importance of ethanoic anhydride and polyesters.

Note:
† Balanced chemical equations are required.
* Conditions required.

TOPIC 16 ORGANIC COMPOUNDS CONTAINING NITROGEN

16.1 Primary amines and amino acids

16.1.1 Formation of primary aliphatic and aromatic amines.

16.1.2 Amines as bases. Ethanoylation using ethanoyl chloride.

16.1.3 Comparative reactions of primary aliphatic and aromatic amines with nitric(III) acid (nitrous acid), HNO_2. Coupling of diazonium salts with phenols, e.g. naphthalen-2-ol and aromatic amines. Azo dyes.

16.1.4 Formulae, structure and amphoteric nature of α-amino acids.

16.1.5 Dipeptides. Outline of protein structure.

16.1.6 Biological aspects. Enzymes.

Learning outcomes Sub-topic 16.1

Candidates should be able to:

(a) describe the preparation of primary aliphatic and aromatic amines from halogenoalkanes and nitrobenzenes respectively;

(b) recall that, and explain why, amines are basic;

(c) recall the ethanoylation reaction of primary amines using ethanoyl chloride †;

(d) compare the reaction of primary aliphatic and aromatic amines with cold nitric(III) acid (nitrous acid), describe the coupling of benzenediazonium salts with phenols such as naphthalen-2-ol and aromatic amines and the importance of this reaction for azo dyes; recall the role of the $-N=N-$ chromophore in azo dyes and be aware that this group links two aromatic rings (cf Topic 12);

(e) recall the general formulae of α-amino acids and discuss their amphoteric and zwitterionic nature;

(f) write down the possible dipeptides formed from two different α-amino acids;

(g) understand the formation of polypeptides and proteins and have an outline understanding of protein structure;

(h) show an awareness of the importance of proteins in living systems, e.g. as enzymes.

Note:
† Balanced chemical equations are required.

16.2 Amides and nitriles

16.2.1 Conversion of acids to amides and the hydrolysis of amides.

16.2.2 Reduction of nitriles using $LiAlH_4$, and the hydrolysis of nitriles.

16.2.3 Industrial importance of polyamides.

Learning outcomes   **Sub-topic 16.2**

Candidates should be able to:

(a) recall the following listed processes and apply knowledge of them to the elucidation of organic problems:

    (i) methods of converting carboxylic acids to amides;
    (ii) the reduction of nitriles with $LiAlH_4$ and the hydrolysis of nitriles and amides;

(b) recall in outline the mode of: the synthesis and the industrial importance of polyamides and understand the similarity of the $-\underset{\underset{O}{\|}}{\overset{H}{\underset{|}{N}}} - C -$ linkage therein to that in naturally occurring proteins.

TOPIC 17   ORGANIC SYNTHESIS AND ANALYSIS

17.1 Calculation of empirical formulae from elemental compositions and derivation of molecular formulae.

17.2 Use of simple mass spectral fragmentation patterns in structure elucidation.

17.3 Use of characteristic infrared frequencies (vibrational spectra) in structure elucidation.

17.4 Use of characteristic n.m.r. shifts and splitting patterns.

17.5 Combination of reactions named in the syllabus to carry out organic conversions.

17.6 Yields in preparative processes.

17.7 Practical techniques - safe procedures.

17.8 Syntheses of industrial and pharmaceutical importance.

Learning outcomes　　　　　　　　　　　　　　　　　　　　　　　　　　　　**Topic 17**

Candidates should be able to:

(a) derive empirical formulae from elemental composition data and deduce molecular formulae from these results plus additional data such as titration values, gas volumes, mass spectrometric molecular ion values and gravimetric results;

(b) use given mass spectral data to elucidate the structure of simple non-cyclic organic molecules (up to and including C_5 molecules, with one chlorine atom) (c.f. Topic 2.2.);

(c) interpret given simple infrared spectra using characteristic group frequencies (supplied in cm^{-1}) : O–H (str), N–H (str), C≡N (str), C = O (str) and N–H (bend)   [str = stretch];

(d) understand that n.m.r. spectra can give information regarding the environment and number of equivalent hydrogen atoms in organic molecules and use such supplied information in structure elucidation *;

* *Candidates will be supplied with simplified n.m.r. spectra of relevant compounds and with a table listing the approximate positions of commonly encountered resonances. They will also be supplied with an indication of the relative peak areas of each resonance and with a note that the splitting of any resonance into n components indicates the presence of n-1 hydrogen atoms on the **adjacent** carbon, nitrogen or oxygen atoms.*

(e) outline the general reaction conditions and basic techniques of manipulation, separation and purification used in organic chemistry, and recall the essential safety requirements during these operations;

(f) propose sequential organic conversions by combining a maximum of three reactions in the syllabus;

(g) deduce percentage yields in preparative processes;

(h) show understanding of the wide applicability of spectroscopic techniques to analytical problems in industry, medicine and the environment;

(i) understand and be able to explain and exemplify the distinction between condensation polymerisation and addition polymerisation (Topic 8, l.o.(i));

(j) recall, as examples of important industrial and pharmaceutical processes, the outline chemistry of the manufacture of polyesters, polyamides and aspirin as rehearsed below:

Condensation Polymerisation (example) – polyesters

$$HOOC\text{-}\sim\sim\sim\text{-}COOH + 2HOCH_2CH_2OH$$

$$\downarrow$$

$$HOCH_2CH_2OOC\text{-}\sim\sim\sim\text{-}COOCH_2CH_2OH + 2H_2O$$

$$\downarrow \text{catalyst}$$

$$\text{-(}CO\text{-}\sim\sim\sim\text{-}COOCH_2CH_2O\text{)}_n$$

The process is carried out at relatively low pressure and 260 °C.

Polyesters are formed from carboxylic acids and alcohols which both possess two functional groups.

Condensation Polymerisation (example) – polyamide

hexanedioic acid → hexane-1,6-diamine

$COOH$ ... $COOH$ →($2NH_3$)→ $COO^-NH_4^+$ ... $COO^-NH_4^+$ →($-4H_2O$)→ CN ... CN →($4H_2$)→ CH_2NH_2 ... CH_2NH_2

$[CONH(CH_2)_6NHCO(CH_2)_4]_n$

nylon 6.6

Polyamides are formed from carboxylic acids and amines which both possess two functional groups

Pharmaceutical Process (example) – Aspirin

Sodium phenate (from phenol and NaOH is reacted with CO_2 under pressure at 100°C and the product acidified to give 2-hydroxybenzenecarboxylic acid (salicylic acid). This is treated with ethanoic anhydride at 90°C to give ethanoyloxybenzenecarboxylic acid (aspirin).

Topic 12   Spectroscopy

In approaching this topic it is useful first of all to make clear some basic principles about the nature of spectroscopy itself. Perhaps the most important of these is to understand that all systems, whether atomic or molecular, can exist in a wide range of energy states, the lowest energy of which is known as the ground state. Under ordinary conditions – at ambient temperatures and in the absence of magnetic fields or electromagnetic radiation – such systems will exist predominantly in their ground states, but when suitably acted upon, most particularly by electromagnetic radiation, may be promoted into various higher energy, excited states. Thus a ray of light, of frequency v, corresponds to a stream of photons, each of energy hv, where h is Planck's constant, and in order for a system in its ground state, of energy E_0, to be excited to a state of higher energy, say E_1, then the incident photon must be of such a frequency that $E_1 - E_0 = hv$. Thus if the frequency of light, v, and therefore the photon's energy, hv, does <u>not</u> correspond to an excitation energy of the system, the photon will <u>not</u> be absorbed and no transition from the ground state to a higher energy state will occur. Consequently, energy is absorbed (or emitted) by atoms or molecules only in certain discrete amounts, or quanta, which correspond to precise changes in the energy of the system concerned, and such energy transfer is said to be <u>quantised</u>, with the energy quantum corresponding to the energy *difference* between one energy state of the system and another. (In principle transitions between one excited state and another are possible, as well as between the ground state and an excited state.) It is most important that at the very outset students should grasp the concept that a spectroscopic transition corresponds to the *difference* in energy *between* one state of a system and another, and <u>NOT</u> to an energy level itself.

The study of spectroscopy thus gives information about the spacing (<u>energy differences</u>) between the various energy levels of a system. In principle such a study can be carried out in two ways, either by exposing a sample to radiation of various frequencies and determining which are (and are not) absorbed (absorption spectroscopy) or by studying the decay of excited states of the system and determining the frequencies of the light emitted (emission spectroscopy).

For atomic systems one is of course concerned with transitions in which electrons are excited to higher energy levels and here attention is directed towards the simplest possible situation, the one electron hydrogen atom, in which the ground state and the higher excited states may all be characterised in terms of appropriate values of the principal quantum number, n. Such systems are usually studied in emission, the electronically excited states being generated either by high temperatures or by high voltage electric discharges.

For molecules however the situation is rather more complex. As before such systems may be electronically excited whereby the distribution of electrons in the various orbitals of the molecule is altered, but molecular species may also be excited by changes in the vibrations of various bonds within the molecule or by changes of rotational speeds about various axes in the system. The latter – rotational excitation – is associated with only very small energy changes and is not here treated further, but the energy differences arising from vibrational excitations are larger, although still smaller than those accompanying electronic transitions. In general molecular systems are usually studied by absorption spectroscopy, with the bands associated with vibrational excitations occurring in the infrared region and those deriving from the higher energy electronic excitations in the visible and ultraviolet regions of the spectrum.

A further complication which arises lies in the fact that instruments used in the study of infrared spectroscopy are usually calibrated either in terms of the frequency, v, or the wave number (see below), which is directly proportional to it, both being directly proportional to energy (via $E = hv$), whilst instruments operating in the ultra violet and visible regions of the spectrum usually work in wave length, λ, which is <u>inversely</u> proportional to energy. It is therefore helpful here first to survey the range of the electromagnetic spectrum and then to establish some points of reference to relate the frequency and wave length scales.

The range of the electromagnetic spectrum of interest to chemists and physicists is usually taken as extending from frequencies of about 10^5, at the low energy end, to about 10^{20}, at the high energy end, the unit of measurement being s^{-1} or Hz (Hertz). The frequency and the wavelength, λ, are of course related by $c = v\lambda$, where c is the velocity of light, $= 2.99 \times 10^8$ m s^{-1}, so that the wavelength can be expressed in m or in some multiple or sub-multiple thereof, the most often used of these being: centimetre, 10^{-2} m; millimetre, 10^{-3} m; micron, 10^{-6} m; nanometre, 10^{-9} m; picometre, 10^{-12} m.

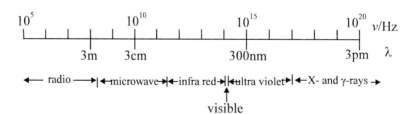

A representation of the electromagnetic spectrum is shown above in which the visible region extends from just over 400 nm at the violet end to about 700 nm at the red end, the nanometre, nm, being the usual wavelength unit used for the ultraviolet and visible region.

For infrared spectroscopy it is customary to express frequencies in terms of a unit $\bar{v}$, the wave number, which is the reciprocal of the wave length ($1/\lambda$) and thus is equal to v/c, where c is the velocity of light. Thus, for example, a frequency of 6.0×10^{13} Hz would, taking $c \approx 3.00 \times 10^{10}$ cm s^{-1}, correspond to a wave number of $6.00 \times 10^{13}/3.00 \times 10^{10}$ = 2000cm^{-1}, the cm^{-1} being the most commonly used unit here. Most infrared spectrometers are calibrated to cover the range between about 700 and 4000cm^{-1}, that between 1500 and 4000 cm^{-1} being of most diagnostic value for organic chemistry (see later). Spectrometers covering the visible and ultraviolet region of the spectrum are usually calibrated to cover between about 1000 and 200nm, corresponding to 10,000 to 50,000cm^{-1} in wave number units. Remember always that both v and $\bar{v}$ are linear in energy whilst λ is reciprocal in energy, by virtue of $E = hv$, and note finally that the general principles of the foregoing introductory material should be kept in mind to achieve full mastery of learning outcomes (a) – (g) below, as well as for better understanding of the remaining outcomes, (h) – (l).

✓ (a) Students should be able to recall the energy gradation across the electromagnetic spectrum, as exemplified by the schematic representation in the above introduction. They should understand where in the spectrum the main regions, including the infrared, visible and ultraviolet regions, occur, and be familiar with either frequency (v) or wavelength (λ) being used to characterise them. In particular they should be quite clear as to which is which as regards the high energy and low energy ends of the spectrum.

✓ (b) Students should be aware that spectroscopic transitions (in either emission or absorption) represent an energy transfer (either from or to the system in question), so that the system passes from one energy state to another. Such a transition therefore represents an energy difference between the two such states of the system. Moreover, such an energy transfer is quantised, that is it can only take place in certain discrete amounts representing such energy differences, and not in any amount – otherwise a continuum (or continuous spectrum) would occur and not a spectrum containing definite peaks and troughs. The study of spectra thus enables these differences in energy between various states of the given system to be determined.

(c) Students should understand that in atomic systems the various accessible energy states correspond <u>only</u> to electronic excitations, i.e. that in the ground state the electron(s) of the atom usually occupy the lowest energy orbitals of those available, whilst in the excited states one (or more) electrons are promoted to higher energy orbitals. These orbitals are of course those characterised by the familiar principal (n) and azimuthal (l) quantum numbers (and in some circumstances also by the magnetic, m, quantum number), but in the special case of the one-electron hydrogen atom system (see (f), below), only the principal (n) quantum number is significant. Essentially therefore in atomic systems spectroscopic excitations come about as a result of the movement of an electron (or electrons) from lower to higher energy levels or vice-versa.

For molecular systems however either electronic or vibrational excitations (or, in principle, both) may be considered. The former correspond, as for atoms, to the movement of electron(s) from one orbital to another, but vibrational excitations correspond to the movement to some extent of the atomic nuclei, rather than of the electrons. In practice the energies associated with vibrational excitations are appreciably smaller than those due to electronic excitations – typically the former will be of the order of about 500 to 4,000 cm^{-1} and the latter in the region of about 5,000 to 50,000 cm^{-1}, but there is no hard and fast boundary between the two types which can well overlap in magnitude. Such vibrational excitations can represent either stretching vibrations (change in bond length) or bending vibrations (change in bond angle), depending on the species involved. Changes in rotational characteristics represent even smaller energy changes – typically in the microwave region – and are therefore not treated in the syllabus.

(d) Students should appreciate that spectroscopy can, in principle, be carried out either in emission or in absorption. In practice, atomic spectroscopy is almost always carried out in emission, in which sharp, clear cut bands are observed, due to only electronic excitations being involved. In this technique the system is excited into a high energy state and the observed emission spectrum represents the energy emitted on the decay of this state to other lower lying energy states. For molecular systems however electronic excitations are almost always accompanied by vibrational excitations, thus leading to substantial band broadening, and this situation is exacerbated when measurements are made in solution, as is often necessary. Thus, unless crystalline solids are available, and can be studied with very high resolution instruments, molecular systems are most often studied by absorption spectroscopy. Here, effectively, the sample is scanned across the whole frequency (or wavelength) range covered by the instrument and what is detected is which frequencies (wavelengths) are absorbed by the sample in being excited from its ground state into various excited states.

(e) The theory of nuclear magnetic resonance (n.m.r.) is complicated and only a very brief outline is given here. (The emphasis here is on <u>application</u> but it is desirable that students should have some idea of how the phenomenon arises rather than simply to treat n.m.r. spectra as just another 'tin can' artefact.)

The starting point is that in general the energetic equivalence of degenerate energy levels (i.e. levels of equal energy), for example atomic orbital levels characterised by the magnetic quantum number, m, (Topic 12(c)) or by the spin quantum number, s, (Topic 1(l)) may be lifted by the application of a magnetic field. Furthermore, just as electron spin can give rise to a magnetic moment which interacts with an applied magnetic field, certain nuclei possess <u>nuclear spin</u> and thus exhibit a magnetic moment which behaves in a similar fashion.

Thus, whilst some common nuclei such as ^{12}C, ^{16}O and ^{32}S have zero nuclear spin (and thus no nuclear magnetic moment), there are many others, such as ^{1}H, ^{13}C and ^{19}F, which do possess such a moment and can therefore interact with a magnetic field. These nuclei all possess a nuclear spin, $I = \frac{1}{2}$, and thus give rise to a two-fold degenerate level which is split by an applied magnetic field. Transitions from the lower to the higher of these states can thus be brought about by the application of radio frequency energy (Topic 12(a) and introduction) and it is the absorption of this energy which is observed in the n.m.r. phenomenon. (Note here that students are only expected to deal with proton, ^{1}H, resonances.)

The n.m.r. experiment requires the use of very strong magnetic fields, at least 10,000 to 20,000 gauss and these correspond to resonant frequencies in the 40 to 100 MHz (4×10^7 to 1×10^8 Hz) region. The experiment is then usually carried out by varying the magnetic field, using a fixed radio frequency, but the alternative technique of using a fixed field and varying the radio frequency can also be employed. Since the position of any n.m.r. absorption (in terms of the magnetic field applied) is strongly dependent on the environment of the ^{1}H atoms involved, this is usually measured in terms of a quantity called the chemical shift, δ, defined as

$$\delta = \frac{H_{sample} - H_{reference}}{H_{reference}} \times 10^6 \, ppm$$

where H is the field strength. The measurement is normally carried out in suitable deuterated solvents (to avoid extraneous proton resonances) and tetramethylsilane, $Si(CH_3)_4$ (TMS), is used as the reference standard since very few species show resonances at higher field strengths than TMS.

(See further Topic 17(d) for required applications of the technique.)

(f) Outcomes (f) and (h) of Topic 12 require that students should (1) understand and be able to describe and interpret the visible atomic spectrum of the hydrogen atom, and (2) understand the relationship between the frequency of the convergence limit and the ionisation energy of the hydrogen atom.

These require respectively familiarity with the Balmer series, based on emission lines terminating on $n = 2$, and a similar knowledge of the Lyman series, "based on emission lines terminating on $n = 1$." In the text relating to outcomes (f) and (h) rather more information than this is given, but this additional material is to be regarded as background information, and students are not expected to deal with series other than the Lyman or Balmer series, or to be able to recall the Rydberg equation. They are however expected to understand the nature of the convergence limit and its relationship to the hydrogen atom ionisation energy. Students should also understand and be able to recall and explain the energy diagram given on p.7 in outcome (f).

Students should know that the emission spectrum of atomic hydrogen may be produced by passing an electric discharge through low pressure molecular hydrogen. In this way atoms of hydrogen are formed in a variety of high energy, excited states. In this one-electron situation these excited states are characterised solely by the value of the principal quantum number, n, and here n takes the values 2, 3, 4, ... ∞, the value $n = 1$ corresponding of course to the ground state (lowest energy) of the hydrogen atom. The emission spectrum of the hydrogen atom thus corresponds to the energies given out when the hydrogen atom in various of these excited states decays to yield an atom in a state characterised by a lower value of n, e.g. from $n = 2$ to $n = 1$ etc.. In fact it is found for the hydrogen atom that the energies of the various orbitals are related according to $E \propto -1/n^2$, where n is the principal quantum number, giving an energy level diagram of the form shown on the next page. [It was then found that the wavelengths of the various lines in the hydrogen atom spectrum fell into a number of clearly defined groups, all of which could be fitted by a relationship of the form

$$\frac{1}{\lambda} = R_H \left[\frac{1}{n_1^2} - \frac{1}{n_2^2} \right]$$

where R_H is a constant, known as the Rydberg constant.] Various series (named after their discoverers) have been identified, in the first of which, the Lyman series, $n_1 = 1$ and n_2 ranges from 2, 3, 4, . . . upwards, towards ∞. This represents the lines generated by energy emission by various excited states, n_2 = 2, 3, 4, . . . ∞, decaying down to n_1 = 1. These occur at fairly high energies with wave lengths in the ultraviolet region, but the next series, the Balmer series, arises when excited states with n_2 = 3, 4, 5, . . . ∞ decay down to n_1 = 2, and this gives rise to a series of which four lines (at 410, 434, 486 and 656 nm) lie in the visible region: these respectively correspond to the transitions $n_2 = 6 \to n_1 = 2$, $n_2 = 5 \to n_1 = 2$, $n_2 = 4 \to n_1 = 2$ and $n_2 = 3 \to n_1 = 2$.

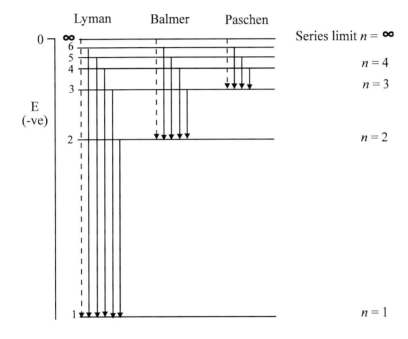

[Various other series have also been identified, with n_1 = 3, 4, 5 and 6, known respectively as the Paschen, Brackett, Pfund and Humphreys series, which represent decay from higher energy levels down to n_1 = 3, 4, 5 and 6, respectively, but these all represent lower energy emissions and thus lie further and further into the infrared region. In all cases however their wave lengths (and hence frequencies and energies) can be represented via the equation given above, taking the Rydberg constant as 1.097×10^7 m^{-1}.]

In the diagram shown above the position of a line from $n_2 = \infty$ to $n_1 = 1, 2$ and 3 respectively has been shown. This clearly cannot be found experimentally, but can be deduced from the known lines by extrapolation, and is known as the <u>series limit</u> or the <u>convergence limit</u>. This is of particular significance for the Lyman series (n_1 = 1), since it measures the energy gap between $n_1 = 1$ and $n_2 = \infty$, or, in other words, the energy required to remove an electron completely from the ground state of the hydrogen atom, and this is treated below in (h).

(g) As noted in the introduction students should be familiar with Planck's law, [$E = h\nu$], and thus appreciate that there is a direct proportionality between energy and frequency. They should also know that frequency and wavelength are related according to $c = \nu\lambda$, where c is the velocity of light, and thus understand the inverse relationship between frequency and wavelength and its consequences.

(h) Students should understand the relationship between the convergence limit of the Lyman series and the ionisation energy of the hydrogen atom, as described in (e), above. [In fact the relationship given in (f) can easily be used to deduce this value. Thus, for $n_2 = \infty \rightarrow n_1 = 1$, there follows:

$$\frac{1}{\lambda} = R_H(1/1^2 - 1/\infty^2) = R_H$$

so that $\quad \lambda = 1/R_H = 1/1.097 \times 10^7 \text{ m}^{-1}$

and $\quad \lambda = 9.116 \times 10^{-8} \text{ m}$

Since $E = h\nu$ and $c = \nu\lambda$ there follows

$$E = hc/\lambda$$

from which one finds

$$E = \frac{6.626 \times 10^{-34} \text{ Js} \times 2.998 \times 10^8 \text{ ms}^{-1}}{9.116 \times 10^{-8} \text{ m}}$$

by substituting the appropriate values of h and c, so that

$$E = 2.179 \times 10^{-18} \text{ J}$$

However, expressed per mole this must be multiplied by the Avogadro Number, so that

$$E = 2.179 \times 10^{-18} \times 6.022 \times 10^{23} \text{ J mol}^{-1}$$

or $\quad E = 1312 \text{ kJ mol}^{-1}$

(The values of ionisation energies are sometimes – especially in older literature – expressed in a unit known as the electron volt, the conversion, requiring multiplication by the factor 1.036×10^{-2} so that $E = 13.59$ eV.). The amount of energy required to remove an electron completely from the hydrogen atom, expressed per mole, is thus quite a large quantity: it is useful to compare this with the amount of energy involved when a molecule absorbs a photon of red light –say at 700 nm, corresponding to the low energy end of the visible range. Since $\nu = c/\lambda$ one has

$$\nu = \frac{2.998 \times 10^8 \text{ ms}^{-1}}{700 \times 10^{-9} \text{ m}} = 4.283 \times 10^{14} \text{ s}^{-1}.$$

Since $E = h\nu$ it follows that

$$E = 6.626 \times 10^{-34} \text{ Js} \times 4.283 \times 10^{14} \text{ s}^{-1}$$

or $\qquad E = 2.838 \times 10^{-19}$ J

Expressed per mole this becomes

$$E = 2.838 \times 10^{-19} \times 6.022 \times 10^{23} \text{ J}$$
$$= 171 \text{ kJ mol}^{-1}$$

which is a not insignificant quantity.]

(i) In emission spectroscopy the emitted light is coloured for wave lengths falling between about 410nm, at the violet end, and about 700nm at the red end (corresponding to some 24,400 and 14,300cm^{-1} respectively in frequency terms). At shorter wavelengths below 410nm (higher energies) the emission lies in the ultraviolet region and at higher wave lengths (lower energies) above 700nm the emission falls in the infrared region. Moreover, throughout these regions there is no essential difference, as far as the spectrometer is concerned, between operating in emission or absorption: in the former case the instrument measures the wavelength (or frequency) of light emitted and in the other case the wavelength (or frequency) of light absorbed.

However, when the human eye is perceiving the colour of substances, either as solids or in solution, the situation is rather different. When, for example, a compound which absorbs at the red end of the spectrum is observed in sunlight (so-called white light), photons at the red end of the spectrum are absorbed and the light reflected from the sample is dominated by those colours which are not absorbed, so that the colour of the compound in white light will be the complementary colour of the light absorbed, in this case blue-green. Similarly, a compound such as an azo dyestuff (see (j) below), which appears orange-red in white light, will appear so to be because it absorbs in the complementary colour region, in this case in the blue, and conversely a substance which appears blue will absorb in the orange-red region. The human eye therefore analyses the colour of the light not absorbed, i.e. reflected, whereas the spectrometer operates by analysing what is absorbed. In emission spectroscopy this problem does not of course arise so that, for example, the four lines of the Balmer series which occur at 410, 434, 486 and 656 nm appear respectively as violet, blue, green and red lines, since the eye perceives the colours actually emitted. (As regards the intensities of colours or of absorptions in other regions of the spectrum, candidates are not expected to have any detailed knowledge and in particular the Beer-Lambert Law is not required, but it should be appreciated that the absorbance of a solution will in general be directly proportional to its concentration.)

To some extent whether or not a compound is coloured is almost accidental, depending on whether or not the energy of the transitions involved falls within the visible region of the spectrum: many species will show transitions of higher energy, corresponding to the ultraviolet region, or lower energy corresponding to the infrared region, but the overwhelming bulk of systems with excitation energies greater than about 5,000cm^{-1} (2,000 nm) will involve electronic excitations, whilst purely vibrational excitations will lie at lower energies, down to a few hundred cm^{-1}. In coloured species therefore the colour will always arise as a result of some electronic excitation and there are numerous examples in both organic and inorganic chemistry. In the latter field however most examples of coloured compounds are to be found in the field of transition metal complexes and thus involve spectroscopic transitions in which the metal *d*-orbitals are involved. (See Topic 21.)

(j) Absorption in the visible and ultraviolet regions (c.f. also (k), below) is often caused by the presence of certain groups of atoms in the molecule, and such groups are known as chromophores, meaning (Greek) 'colour bringers'. (To some extent this term is a misnomer since many functional groups bring about absorption in the ultraviolet region, from which no observable colours will result, but the term is in general use for absorptions in both the visible and the ultraviolet region.) Thus systems containing the carbonyl, >C=O, group have a strong absorption at around 185 nm and those containing the double bond, >C=C<, at about 170 nm, but these lie at shorter wave lengths (higher frequencies and energies) than are usually accessible for ultraviolet measurements (200nm). However, when the >C=C< double bond forms part of a conjugated system of alternating double and single bonds, such as >C=C−C=C−C=C<, the absorption due to the unsaturation moves progressively to longer wave lengths, and if the carbon chain is long enough may even lead to absorption in the visible region of the spectrum. [This comes about because as the chain lengthens more and more orbitals are involved in the delocalised π-electron system. These orbitals in the molecule involve all the carbon atoms of the system – 'molecular orbitals', see below – and as their number increases the energy gap between the ground and excited states of the system diminishes, so that the absorptions correspond to smaller frequencies and longer wave lengths. Thus, for this reason the compound β-carotene, the colouring material in carrots, which has a conjugated chain of 18 carbon atoms, absorbs strongly at around 450 nm in the blue region, and is therefore a bright orange colour (see (i), above).]

In the field of organic chemistry however the most striking example of a chromophore is probably that of the −N=N−, azo, group, especially in aromatic compounds derived from azobenzene. (See especially Topic 16.1 (d), below.) These species are also mostly orange or orange-red in colour, due to absorptions in the blue region, but the presence of substituents in the benzene rings can cause significant shifts in the absorption positions, leading to a wide variety of colours.

Students should therefore understand that the concept of a chromophore involves the idea that a group of atoms or orbitals in a molecule can be responsible for absorptions being produced in a particular region of the spectrum. [Of course, once one is dealing with molecules (rather than atoms) the orbitals involved are really molecular orbitals which in principle involve all the atoms and orbitals in the system.] In many cases however the electronic excitations leading to spectroscopic absorptions can be regarded as largely localised in a particular region or a particular set of orbitals of the molecule, and are usually only slightly influenced by the rest of the molecule. In this way therefore the idea of a chromophore is a useful concept, especially in organic chemistry, although some aspects of transition metal chemistry may also be explored in this way. (See Topic 21.)

(k) In just the same way as that in which the concept of a chromophore was applied to electronic transitions in (j) above, a similar approach may be adopted in characterising vibrational frequencies. Thus, in principle, the vibrations of molecules are complex and involve all the atoms, with the possibilities of <u>both</u> the stretching of bond lengths <u>and</u> the deformation (or bending) of bond angles. In practice however, especially in organic chemistry, it is found that for many groups of atoms certain characteristic vibrations occur at much the same frequencies, irrespective of the rest of the molecule, and thus appear to be essentially localised in particular bonds, and in particular vibrations, either stretching or bending.

Students should therefore appreciate that the determination of the infrared spectrum of a given organic molecule can reveal important information as to the presence or absence of particular groupings within that molecule. For such diagnostic use the most important part of the spectrum is that lying between about 1,500 and 4,000 cm^{-1}, and the most important peaks in that region are given in the chart below. (Note that infrared peaks are sometimes designated in terms of the micron, 10^{-6}m, as a wavelength – see introduction to this Topic, above.)

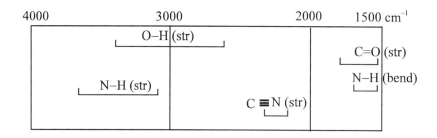

Students should especially note that these characteristic infrared peaks are indicative of the presence of particular vibrations, <u>not</u> of the presence of particular functional groups. Thus, for example, aldehydes (–CHO), ketones (>C=O), carboxylic acids (–COOH), esters (–COOR), acid chlorides (–COCl) and amides (–CONH$_2$) will all show a peak in the 1500 – 1750 cm^{-1} range, due to the C=O stretching vibration. Similarly <u>both</u> amines (–NH$_2$) and amides (–CONH$_2$) will show the stretching and bending vibrations of the N–H fragment and the O–H stretching vibration will be exhibited by both alcohols (–OH) and carboxylic acids (–COOH). The actual positions of these peaks <u>will</u> vary to some extent within the ranges shown in a predictable way depending on the nature of the functional group and its location in the molecule, but a study of this is beyond the scope of the syllabus and not expected of students. Involvement in hydrogen bonding does tend to broaden the peaks due to the O–H stretching vibration, and to a lesser extent those due to N–H stretching, but the peak due to the C=O stretching mode is almost always strong and not too broad, whilst the most characteristic of all is the strong very sharp peak due to C≡N stretching at about 2,250 cm^{-1}.

In general students will be expected to use given infrared observations to determine the presence or absence of particular groupings in an organic molecule. They will also be expected to be able to use such information <u>in conjunction with</u> other data (e.g. analytical data, information about reactions, chirality or otherwise, etc.) to deduce the identities of compounds, interpret reaction sequences, suggest synthetic routes and so on, as well as being able to indicate where they would anticipate infrared peaks to occur in this context. In all cases the chart of frequencies of peaks given above will be supplied: candidates will <u>not</u> be expected to remember such frequency values. (See also Topic 17 (c) below.)

(l) In the majority of cases the use of spectroscopy for analytical purposes is based upon absorption rather than emission spectroscopy. The only simple exception to this lies in the qualitative use of flame tests (see Module CH1, Topic 5.2 (e) and Module CH5, Topic 19 (c)) for the detection of certain elements, mostly those of Group I and Group II of the

Periodic Table. Basically this technique simply requires the detection – by the human eye – of the characteristic colours in the emission spectra of these elements.

Students should therefore recognise that absorption spectroscopy is most commonly used and also that, although techniques are available for the manipulation of solids and gases, most substances are investigated in the liquid phase, usually in solution in a solvent with little or no absorbance in the region of interest. In general the quantitative measurement of absorbance is fairly straight forward in the ultraviolet and visible region over the 200 to 1000 nm range and since absorbance is directly proportional to concentration the latter can thus be readily determined. Moreover, when substances are strongly absorbing at one or more wavelengths, the quantitative measurement of the absorbance permits the measurement of very small amounts of the substance in question to be carried out which is clearly useful in a number of applications in industrial, medical and environmental chemistry.

Students should also be aware that spectroscopic measurements find a wide applicability in kinetic studies since in this way, by studying the variation of absorbance at a given wavelength with time, the rate of increase or decrease of the concentration of a given species may be determined and thus both the rate equation and the appropriate rate constant determined.

In general infrared spectroscopy is less often used quantitatively and the more usual use is in a qualitative sense to ascertain the presence or absence of absorptions characteristic of a particular grouping. (See (k) above.) Most frequently the samples either as liquids or finely powdered solids in a hydrocarbon mull are simply examined for the presence or absence of the characteristic peaks such as those listed in (k) above. Techniques are available for the quantitative measurement of absorbances in the infrared region for solution samples, but are more difficult to use because the intensities of infrared bands (due to vibrational excitations) are almost always markedly smaller than those of ultraviolet and visible bands (due to electronic excitations). The factors governing the intensities of spectroscopic transitions are however too complicated for further treatment here, save only to note that moving the electrons around (electronic) usually leads to more intense bands than moving the nuclei around (vibrational).

Note finally that students are not required to have any knowledge of the mode of construction or operation of any of the spectrometers referred to above or any other technical details relating to the mode of presentation of the relevant spectra: the requirement is that they understand the essential nature of spectroscopy and can explain and interpret such data.

Topic 13   Isomerism and Aromaticity

(a) In addition to the nomenclature requirements rehearsed in Topic 7(a), candidates should now also be able to give the systematic names of all the simple compounds, including benzene derivatives (i.e. those containing phenyl, C_6H_5-, and substituted phenyl groups), which contain any of the functional groups treated in this Module.

(b) – (d) As well as understanding the concept of geometric isomerism, covered in Topic 7(d), as a form of stereo (or spatial) isomerism, students should also appreciate that this latter term embraces a further form of isomerism of which organic chemistry provides a vast number of examples. This is called 'optical isomerism', for reasons to be explained later, and occurs when four <u>different</u> atoms or groups are attached to a single carbon atom – obviously a saturated carbon atom possessing four <u>single</u> bonds. Suppose one has a molecule whose shortened structural formula may be written as $ClCH(CH_3)OH$, or graphically as

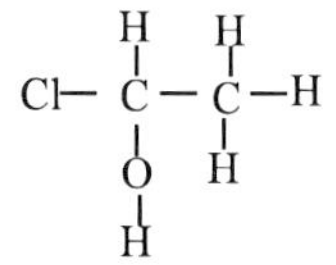

$$Cl-\underset{\underset{H}{|}}{\underset{|}{\overset{H}{\overset{|}{C}}}}-\underset{H}{\overset{H}{\overset{|}{\underset{|}{C}}}}-H$$

This flat, 2-dimensional form does not of course display the true geometric arrangement about the carbon atom and a better representation might be something of the form shown below; in which the ▶ and … symbols represent linkages coming out of and going into the plane of the paper.

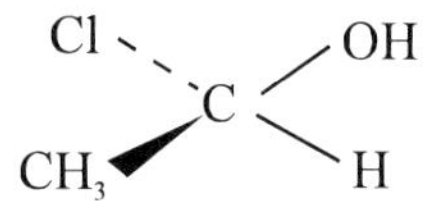

However, an alternative formulation is equally possible, and is shown below,

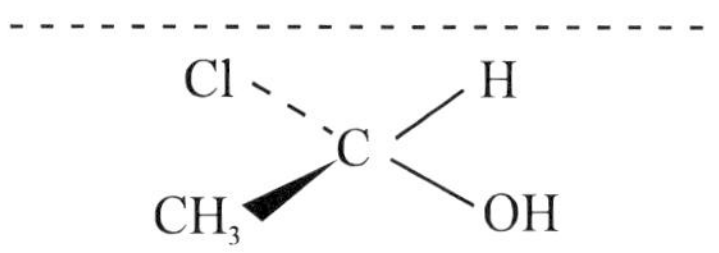

which is in fact the mirror image (across the plane indicated) of the first formulation. Moreover, it is clear that these two formulations of the molecule are not superimposable and therefore not identical: one is in fact a reflection of the other, and they are related as is a left hand to a right hand.

In fact any molecule with four different groups or atoms attached to one carbon atom exhibits the property of existing in both 'left hand' and 'right hand' forms and is said to exhibit 'chirality', such a carbon atom bearing the four different groups being termed a 'chiral centre'.

The two different forms of a compound exhibiting chirality are called 'enantiomers' of each other and usually exhibit similar physical and chemical properties, apart from their effect on polarised light. Thus if a beam of plane-polarised light is passed through solutions containing equal concentrations of the two enantiomers, then in one case the plane of polarisation will be rotated in one direction (say, right) and in the other case in the opposite direction (left) to the same extent. For this reason the phenomenon of chirality is often accorded the alternative description of 'optical isomerism' and such systems are said to show 'optical activity'.

When chiral compounds are produced from non-chiral materials the product is usually a 1:1 mixture of the two enantiomers and will therefore show no resultant rotation of the plane of polarisation of the light. Such a mixture is known as a 'racemic mixture' or a 'racemate'. However, another property of enantiomers is that they react in different ways with other chiral molecules and such a procedure is frequently the basis for the resolution of a racemic mixture, that is for its separation into two enantiomers.

Candidates should therefore understand and be able to explain all the terms used above and also be able to identify chiral centres in given molecules. Apart from S Level papers, candidates will not be expected to deal with more than one chiral centre in a given molecule. Finally it should be noted that chirality (optical activity) is again not restricted to organic systems (although it is far more often observed there), and there is a steadily lengthening list of basically inorganic species which also display this phenomenon.

(e), (f) These two outcomes treat the notion of aromaticity as exhibited in the simplest aromatic hydrocarbon, benzene, C_6H_6. In particular students must appreciate the connection between the alternating double bond, single bond representation and the delocalisation of the π-electron density found in this molecule. Students must thus be able to describe the structure of benzene and to explain the nature of the bonding therein. They should be able to appreciate the conceptual difficulties involved in understanding the lack of reactivity in a molecule which, formally, contains three double bonds. Thus when Kekulé established the cyclic nature of the structure, which could be represented by the two equivalent forms

 and

there was difficulty in reconciling such a formulation with the fact that benzene shows no tendency at all to undergo addition reactions across any of those double bonds.

It must therefore be clearly understood that the structure is that of a regular hexagon with <u>all</u> the bond lengths and bond angles equal, but with a C–C distance (139 pm) intermediate between that expected for a C=C double bond (134 pm) and for a C–C single bond (154 pm). This is most readily interpreted in terms of a carbon framework in which the C–C bonding in the plane of the ring is due to σ– type bonding (in which electron density accumulates between the carbon atoms) and the C–H linkages similarly arise from σ– type bonding. In this way a total of 6 electrons (one from each carbon atom) remains after the σ– framework has been accounted for and these have to be accommodated in orbitals formed (by π– type bonding) from the 6 p-orbitals orientated at right angles to the plane of the ring. These 6 electrons are in turn accommodated in three delocalised orbitals which extend over all the 6 carbon atoms of the ring; this π– bonding thus leads to charge clouds evenly delocalised above and below the plane of the ring. Thus the bonding in benzene leads to a bond order intermediate between a single and a double bond and accounts for the lack of reactivity in addition reactions. It should also be appreciated that this delocalised structure for benzene can be shown <u>experimentally</u> to be significantly more stable than the hypothetical Kekulé structures with alternating double and single bonds.

The calculation of the delocalisation energy of benzene may be made in various ways – using bond energies directly or from hydrogenation, combustion or similar enthalpy data. However, it is the logic underlying the calculations which seems to cause confusion. Basically we find that the energy contained in real benzene, in a bottle, is less than that calculated for a hypothetical molecule, , which contains three single and three double carbon-carbon bonds. This difference is the <u>delocalisation</u> or resonance <u>energy</u>, the decrease in energy or increase in stability conferred when the π electrons spread evenly over the whole ring. Estimates vary but 150kJ is a reasonable value. Note that delocalisation is exothermic so strictly we should write –150kJ mol^{-1}.

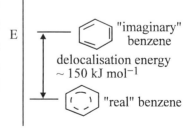

The principle of the methods is to estimate the energy of "imaginary" benzene in some way and compare this value with that of real benzene. The bond energy method calculates $\Delta H_f^\ominus$ of "imaginary" benzene by adding the atomisation enthalpies of 6C and 6H atoms and subtracting the bond energies of all the bonds in "imaginary" benzene (6C–H, 3C–C, 3C=C), – subtracting since bond <u>formation</u> from gaseous atoms is exothermic.

The hydrogenation met<u>hod</u> assumes that ΔH (hydrogenation) of "imaginary" benzene will be three times that of ⬡ and compares this value with that of real benzene, since both benzenes would give cyclohexane (C_6H_{12}) on complete hydrogenation.

(g) It should be appreciated that, by virtue of the structure described above, the benzene ring is an area of high electron density, just as was the double bond in alkenes. (See Topics 7 and 8) It is therefore also susceptible to attack by electrophilic reagents, but, because of the greater stability of the delocalised system in benzene, more robust conditions and more vigorous reagents are needed.

Students should therefore recall that both the nitration and the halogenation reactions of benzene are examples of <u>electrophilic substitution</u>: here an attacking entity (effectively Cl or Br and NO_2) is <u>substituted</u> for a hydrogen atom of the aromatic ring. The syllabus requires knowledge of the mechanism of the nitration reaction, which is outlined below, and that of the halogenation reaction is appended for comparison since it follows a very similar path.

Students should know that the (mono) nitration of benzene is carried out with a mixture of concentrated nitric and concentrated sulphuric acids at <u>about 55°C.</u> These acids react together according to:

$$HNO_3 + 2H_2SO_4 \rightarrow \underline{NO_2^+} + H_3O^+ + 2HSO_4^-$$

the nitronium ion, NO_2^+, being the <u>active electrophile</u> in effecting nitration.

Such an electrophile attacks the $\pi-$ system of the ring, using two electrons therefrom, to produce initially an intermediate (known as a <u>Wheland</u> intermediate, after the propounder of this mechanism) in which the remaining four electrons from the $\pi-$ system of the ring are delocalised over the remaining five carbon atoms.

Thus

⬡ + NO_2^+ → [cyclohexadienyl cation with H and NO_2]

followed by

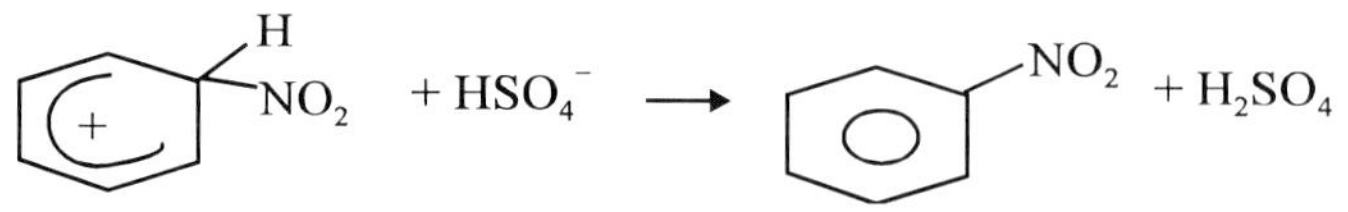

[benzenium intermediate with H and NO₂] + HSO₄⁻ → [benzene with NO₂] + H₂SO₄

Halogenation follows a very similar path but here a so-called halogen carrier such as iron(III) chloride or aluminium chloride is needed to convert the free halogen into a positively charged ion.

Thus, with aluminium chloride as the carrier one may write

$2Cl_2 + Al_2Cl_6 \rightarrow 2Cl^+ + 2AlCl_4^-$

(Here the electron deficient species is essentially AlCl₃ which takes up Cl⁻ from Cl₂, leaving the required positive ion, Cl⁺.)

Then

[benzene] + Cl⁺ → [benzenium intermediate with H and Cl]

followed by

[benzenium intermediate with H and Cl] + AlCl₄⁻ → [chlorobenzene] + AlCl₃ + HCl

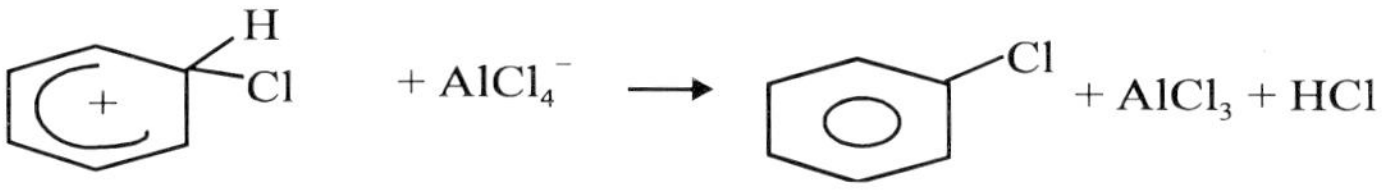

which is again an electrophilic substitution. Essentially the halogen carrier is an electron deficient species which can pull off Hal⁻ from Hal₂ to leave the necessary Hal⁺ ion.

(h) It should be recalled that benzene is appreciably more resistant to electrophilic attack than most alkenes and in marked contrast to alkenes shows no tendency to undergo addition reactions. This should be explained by noting that the delocalised system (represented by benzene as ⌬) is much more stable than the alternating double and single bond structure (represented by benzene as ⌭). This can in fact be shown by comparing the value of the heat of hydrogenation for benzene which is much less negative than would be expected for a system apparently containing three double bonds, i.e. equivalent to three ethenes. Consequently the delocalised system of benzene is markedly the more stable.

It is also advantageous for students to be aware of some of the older classical evidence for the equivalence of all the C–C linkages in benzene, for example the fact that it is not possible to produce isomers of 1,2-disubstituted benzenes which differ in the location of the double bonds with respect to the substitutents.

Thus

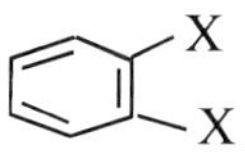

[1,2-disubstituted benzene with X,X] and [1,2-disubstituted benzene with X,X with alternate double bond position]

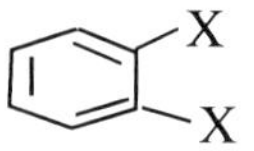

are not different species and such a compound is simply represented as

$$\text{C}_6\text{H}_4\text{X}_2$$

From a nomenclature point of view it is useful to note that the chemistry of benzene and its derivatives is usually known as <u>aromatic</u> chemistry (originally from the fragrant aromas of many such compounds) whereas the chemistry of compounds with various simple hydrocarbon chains which do not contain a benzene ring is known as <u>aliphatic</u> chemistry (from the Greek for fat since the fats contain such simple carbon atom chains). Hydrocarbons such as benzene are of course unsaturated, just as are the alkenes, but hydrocarbons containing a benzene ring are known generically as <u>arenes</u> so as clearly to distinguish them from the alkenes.

Topic 14   Organic Compounds Containing Halogens

These compounds are important in many ways today and some awareness of this importance and associated problems is required under outcomes (d), (e) and (f), but the actual chemical content is fairly small and mostly restricted to chloro-compounds. However this is a good area in which to appreciate the changes brought about by substituting a functional group for hydrogen in an alkane. Four points may be made.

1. Bond strength

C–H	(C–F)	✳	C–Cl	C–Br	✳	(C–I	
412 kJ	484 kJ		338 kJ	276 kJ		238 kJ	mol⁻¹)

In general the greater the bond strength the lower the reactivity so that fluorocarbons are less reactive than alkanes and the other haloalkanes increasingly more reactive and less stable from Cl → I.

2. Bond polarity. This increases from the covalent C–I bond to the highly polar C–F bond (see electronegativity differences). In the case of C–Cl and C–Br bond reactions which are the only ones included in the syllabus, the combination of not very high bond strength and polarity means that bond breaking may occur heterolytically as in (b) or homolytically as seen in the problems of CFCs in (d) and (e).

3. Van der Waals' forces. Halogenoalkanes exhibit relatively strong intermolecular forces, owing to the large number of electrons in the halogen atoms, and tend to be <u>liquids</u> having a good solvent power for organic compounds, although not soluble in

( ) not on syllabus

water to any extent.* The ability to dissolve fat is a reason for their toxicity to humans, where damage to liver or kidneys may occur. * *related to solvent abuse, LFC*

4. Flammability. Alkanes are highly flammable since both the carbon and hydrogen oxidise exothermically to stable molecules. Oxides of the halogens are unstable and not formed exothermically, so that halogenoalkanes become increasingly non-flammable as hydrogen is replaced by halogen.

Finally, in this introduction the following points may be made to clarify the extent of the syllabus in this topic.

1. The halogenoalkanes are confined to those of Cl and Br, plus the knowledge that C–F bonds are strong and the awareness in (d) and (f).

2. Knowledge of the effects of the structure of the <u>alkyl</u> group on the course of any reaction is <u>not</u> required, but the content of (c) must of course be known.

3. Candidates must know that both substitution and elimination (Topic 7(g)) reactions may occur and know which reaction condition favours which route, but <u>need</u> <u>not</u> know the effect of structural factors on the preferred route (i.e. as stated in 2.).

4. The terminology S_N1 and S_N2 is <u>not</u> required; outcome (b)(i) is limited to primary halogenoalkanes and <u>no</u> knowledge of the hydrolysis of tertiary halogenoalkanes is expected; (see however below).

Some specific points now follow.

(a) The direct chlorination is closely tied in with Topic 8(d), where methane is a model compound for the process, which is used with various alkanes, all being radical in type. Radical reactions are random in nature so that any of the C–H bonds may be substituted and mixtures of halogenoalkanes obtained. This may not matter if, e.g. the product is used as a solvent, and can be controlled to some extent by adjusting the chlorine/alkane ratio. Initiation of the chain process by breaking the Cl–Cl bond may be performed by either light or heat energy. – *UV light*

As a non-syllabus comment the industrial situation is complex, with chlorination of alkanes and alkenes, oxychlorination, and the manufacture of both chloroalkanes and chloroalkenes being important.

(b) This is mainly straightforward textbook material involving nucleophilic substitution but (i) provides the main link between kinetics and mechanism so far as the syllabus is concerned. Kinetic measurements show that the hydrolysis is first order with respect to both 1–bromobutane and OH^-, so that both of these species must be involved in the rate-determining step – one molecule of each. There is a tendency to think that this must always be the case in a reaction such as $RBr + OH^- \rightarrow ROH + Br^-$, but this is not so; we have to experiment to find the truth. Although the hydrolysis of tertiary halogenoalkanes is not in the syllabus, it provides useful background knowledge which helps us to obtain a firm understanding of kinetics and mechanism, since this reaction, which on paper is identical to the 1–bromobutane hydrolysis, obeys first-order kinetics overall, with the rate being quite independent of the OH^- concentration. Thus despite the balanced equations being of the same form, the two mechanisms are different, and this fact can only be established by measurements of rates at different concentrations.

A point to note about the substitution reaction with CN^- is that this adds a covalently bonded carbon to the alkyl chain and is thus a valuable means of increasing chain length.

(c) The much more severe conditions needed to hydrolyse chlorobenzene compared with a chloroalkane (350°C as against 100°C) result from the greater strength of the C–Cl bond when attached to an aromatic ring; i.e. 400 kJ mol^{-1}, as against 350 kJ mol^{-1}. Thus more energy is needed to break the bond and fewer of the molecules will possess the higher activation energy needed for reaction. (The activation energy will of course be much less than 400 kJ since bonds are being formed as well as broken in the transition state.)

The greater C–Cl bond strength in chlorobenzene is believed to result from overlap between p electrons on the chlorine and the delocalised p (π) electrons in the ring to give some π character to the bond and thus strengthen it.

(d), (e) The good solvent power of halogenoalkanes for organic materials has been discussed above. This power varies with the type of compound: chlorocompounds are usually strong solvents (CH_2Cl_2 is used in paint strippers), fluorocarbons are less good, and chlorofluorocarbons intermediate in solvent power, e.g. Cl_2FC-CF_2Cl, or $CFCl_3$, which is widely used for degreasing and dry cleaning.

Refrigerant and aerosol uses developed from the fact that CFCs are inert, non-flammable, and often gases which could be liquefied at room temperature by the application of moderate pressures.

Essentially the same properties are needed in anaesthetics and Halothane, $CF_3CHBrCl$, is widely used; in a somewhat different medical use, freezing sprays for temporary pain relief of sprains etc use the evaporative cooling of chloroethane.

The usefulness of halogenoalkanes led to production on a large scale but two adverse effects emerged, the problem of solvent abuse and toxicity, and the environmental effects of CFCs. The effect of taking in powerful fat solvents on the liver and kidneys have already been referred to. Solvent abusers may become unconscious due to the anaesthetic effects, which again may relate to fat solubility. The ion channels which operate nerve impulses are controlled by fatty ion "gates" which cease to operate when affected by halogenoalkanes.

In the case of the environmental effects the very inertness of the compounds has meant that they persist in the atmosphere (half-lives of up to 100 years), and diffuse into the upper atmosphere, or stratosphere, where high energy ultraviolet radiation from the Sun, which does not penetrate to lower levels, breaks carbon-halogen bonds to form radicals, such as chlorine atoms. These then catalyse the destruction of ozone through cycles such as

$$\begin{array}{r}Cl\cdot + O_3 \longrightarrow ClO\cdot + O_2 \\ ClO\cdot + O \longrightarrow Cl\cdot + O_2 \\ \hline O + O_3 \longrightarrow 2\,O_2\end{array}$$

One chlorine atom may thus destroy many ozone molecules. Thus the ozone layer, which protects us from the damaging, e.g. cancer-producing, effects of short wavelength ultraviolet radiation, is lowered in concentration, and, under certain conditions, such as around Antarctica in winter, may disappear altogether, leaving the Earth's surface fully exposed.

The very strong C–F bonds are not broken so that a typical reaction is

$CF_2Cl_2 + U.V. \rightarrow CF_2Cl\cdot + Cl\cdot$

Weaker C–Br bonds will break even more readily than C–Cl so that the relative ability of a compound to remove ozone depends upon the compound, e.g. $CBrClF_2$ is ten times worse than CCl_3F.

Realisation of the seriousness of the problem has led to an international agreement to reduce CFC emission.

(f) Organochlorine compounds, especially aromatic or ring compounds containing several chlorine atoms (DDT, BHC, aldrin, dieldrin) have been widely used in the ceaseless war against insect pests which spread disease, and have saved an estimated 4 million deaths per year from malaria alone as well as preventing many crop failures caused by plagues of insects. However these tend again to be both stable and fat-soluble and concentrate in the food chain, finally killing birds or stopping successful breeding, and are being replaced by organophosphates.

Organohalogen polymers may be represented by PVC (polyvinylchloride) – 17 million tonnes per annum worldwide, and PTFE (polytetrafluoroethene) – 25 thousand tonnes per annum, although there are several others. PVC – strictly polychloroethene – is made in rigid form for pipes, gutters, window frames, bottles etc., and in flexible form for

raincoats, Wellington boots, coatings, and electrical insulation. The polar C–Cl bond increases intermolecular forces between the polymer chains and thus polymer rigidity.
PTFE is fully fluorinated and is thus chemically resistant, non-flammable and <u>thermally stable (strong C–F bond)</u>. It is used for bearings, gaskets, and resistant films. The complete cover of fluorine atoms on the polymer surface leads to very low friction and thus use in <u>non-stick coatings</u> for, e.g., <u>frying pans</u> (also heat resistance); PTFE is an excellent electrical insulator for similar reasons.

N.B. Comments on outcomes (d), (e) and (f) are in no way meant to define a syllabus, but rather to illustrate the range and importance of these topics today and some of the key questions surrounding the use of halogeno-compounds. Awareness might clearly be shown in other ways using other examples.

Topic 15   Organic Compounds Containing Oxygen

Topic 15.1   Alcohols and Phenol

This topic causes few difficulties and has some straightforward chemistry plus the wider aspects of the large scale use of ethanol as an industrial chemical, its importance in society as a social drink, its use as a fuel, and the historical shift between fermentation and ethene hydration as preferred routes of manufacture. Industrial production is measured in millions of tons per annum, with major uses including the manufacture of esters and as a solvent.

(a) The physical properties of the lower alcohols are dominated by the effects of the polar O–H group, and especially hydrogen bonding between water and alcohol. Also boiling temperatures are comparatively high (compare ethanol, b.t. 77°C with propane, –42°C, both having similar R.M.M. values) since hydrogen bonding between alcohol molecules means that the effective R.M.M. is much larger than 46.
The dominance of the O–H group decreases as the alkyl chain length increases so that solubility in water decreases rapidly from C_4 onwards and higher alcohols become increasingly oily or waxy and hydrocarbon-like.

(b) These straightforward methods are not of course used industrially but for laboratory and research purposes. One important general point to remember, which is relevant here,

RMM – relative molecular mass

is that we get two for the price of one with many of the outcomes in organic chemistry. Thus here, forming alcohols from halogenoalkanes is exactly the same reaction as in Topic 14(b)(i), and, from carbonyl compounds the same as Topic 15.2(c). Consequently, interconversion charts such as that shown below provide a valuable way of both learning and integrating the factual base of organic chemistry.

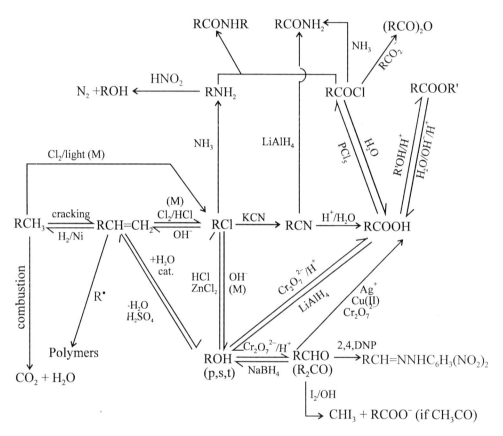

(M) denotes a syllabus mechanism

(c) What is required here is to know that ethanol is made industrially on a large scale from ethene and steam by passing over a heated acid catalyst (see also Topic 8(g)). No technical details of the process are expected. As a matter of background the sulphuric route has been largely replaced by the direct use of a solid-supported phosphoric acid catalyst. Also as background note that this is an <u>electrophilic addition of H⁺</u> (compare with 8(f), HBr addition).

(d) Once again, as in (b) we are dealing mainly with <u>reversible systems,</u> i.e.
  1. ROH + HBr ⇌ RBr + HOH   (Topic 14(b))
  2. ROH + R'COOH ⇌ ROCOR' + HOH   (Topic 15.3(b))
  3. CH₃CH₂OH ⇌ CH₂=CH₂ + HOH   (Topic 8(g)),

and control the position of equilibrium in a Le Chatelier way by concentration and reagent control. Thus in 1. the use of <u>anhydrous HBr (KBr + H₂SO₄)</u> gives a <u>good yield</u> of <u>bromo-compound</u>, whereas <u>excess aqueous alkali</u> with the bromide gives the alcohol.
In 2. the use of <u>concentrated H₂SO₄ removes water</u>, as well as <u>acting as a catalyst</u>, to increase ester yield; the direction is reversed by the use of <u>excess water</u> and <u>acid or alkali.</u>
In 3. again either the left or right hand side may be made to be dominant by adjusting conditions.
Ethanoyl chloride provides an effective means of converting alcohols into esters since the only other product is HCl which is absorbed by a basic solvent such as pyridine.

(e) This outcome is clearly linked with outcomes (a), (b), and (c) of Topic 15.2, the main points of interest being the difference in behaviour between primary and secondary alcohols and the practical effects of the instability to oxidation of the initial, aldehyde, product in the case of primary alcohols.
The aldehydes from primary alcohols still possess an oxidisable hydrogen attached to the carbonyl group and will therefore convert to carboxylic acids unless removed from the system. Ketones from secondary alcohols may only be oxidised further through C–C bond rupture and are therefore resistant. Acidified manganate(VII) ion ($E^\ominus$ = 1.52V) oxidises primary alcohols directly to acids, whereas the slightly weaker acidified dichromate(VI) ion ($E^\ominus$ = 1.33V) first forms an aldehyde, which may be removed.
Note in connection with (f) that the ethanol reduction of <u>orange dichromate(VI) to green chromium(III)</u> was the basis of an earlier <u>breathalyser test based on colour change</u>, but the voltage developed by an ethanol-oxidising fuel cell is now largely preferred as an indicator of "being over the top".

(f) This is an awareness topic of which some candidates may have considerable personal experience. The current medical view is that a <u>moderate</u> intake of ethanol-containing drinks may be beneficial to health, but the dangers of excess to one's own psychological and physical (liver, kidneys, brain) health and to one's fellow human beings remain.

The drinks, consisting essentially of more or less dilute aqueous ethanol are prepared by the yeast fermentation of sugar. This is a very ancient (3×10^9 years) internal redox process which the yeast plant uses an an energy source:

$$C_6H_{12}O_6 \rightarrow 2C_2H_5OH + 2CO_2$$

oxidation no   (0)   (–II)   +IV
of carbon

Once the ethanol concentration reaches 90g dm^{-3} (say about 12%) the yeast die so that while the weaker drinks – beer and alcoholic lemonade (5%), and wine (10%) may be produced directly, distillation is needed to obtain spirits (30% +).

(g) The concept of using alcohols as a fuel, especially in motor cars, has been around almost as long as the car itself, and has had many ups and downs. <u>Methanol</u> has been used as a <u>racing car fuel</u> and ethanol added to petrol. Lower enthalpies of combustion (44 kJ per gram for petrol against 19 kJ per gram for alcohols) favour hydrocarbons – alcohols are partly oxidised hydrocarbons – but the major recent impetus has been the <u>fear of oil running out</u>, while ethanol from the <u>fermentation of biomass</u> comes from a <u>renewable resource.</u>

Ethanol as a motor fuel has made some headway in tropical countries such as Brazil and Kenya, which are suitable for sugar cane growth and have limited oil reserves. Disadvantages are that fermentation is slow and produces a dilute solution of ethanol, which must be distilled. The process is unlikely to become of major importance generally unless there is a steep rise in the cost of oil. ✱Obviously, in general terms, burning industrially produced ethanol does not make much sense after carrying out the steps alkane → alkene → ethanol, when the alkane may be burnt directly and give more energy. ✱ *interesting!*

(h) The increased acidity of phenol compared with ethanol results from the <u>stabilisation of the phenoxide ion</u> through <u>delocalisation</u> between the <u>oxygen p orbital</u> and the <u>p (π) benzene ring orbitals</u>, which <u>spreads the negative charge</u>.✱ Thus the equilibrium $C_6H_5OH \rightleftharpoons C_6H_5O^- + H^+$, is displaced in the direction of acid dissociation and phenol is noticeably acid, although only mildly so ($K_a = 10^{-10}$).

✱ similar to carbocation, tertiary is more stable because of spread of +ve charge

The same effect is reflected in the ready reaction of bromine with the ring, electrophilic substitution being promoted by the fact that the ring is more negative than in benzene.

(i) Aqueous iron(III) chloride solution is a very useful reagent for the detection of phenolic – OH groups in aromatic (benzenoid) compounds. Very many (although not quite all) such compounds give intense colours with this reagent, most of which are mauve to deep violet, although in some cases green or red colourations are produced. Alcohols do not give this reaction.

Topic 15.2   Aldehydes and Ketones

Although we now have the appearance of the reactive carbonyl group, the syllabus focuses mainly on oxidation and reduction and on the chemistry relating to distinguishing between aldehydes and ketones and their identification in analysis. The properties of these compounds often appear in problems on organic transformations since they occupy an intermediate position with respect to the degree of oxidation.

As non-syllabus background methanal, ethanal and propanone are important industrially at the million tonne level as solvents (propanone), intermediates (ethanal for ethanoic acid) and for resins and plastics (all three).

Also as background is the fact that the typical addition to the polar C=O double bond involves a nucleophilic attack on the positive carbon, unlike the electrophilic addition to alkenes; however this type of reaction only appears in outcomes (d) and (e) and no further mechanistic understanding is required.

Primary alcohols, RCH_2OH, may be oxidised by warming with acidified potassium dichromate(VII), $K_2Cr_2O_7$, first to aldehydes, RCHO, and then to carboxylic acids, RCOOH, whilst secondary alcohols, RCH(OH)R′, on similar treatment, yield ketones, RCOR′. Both aldehydes and ketones contain a free carbonyl group, >C=O, and undergo numerous characteristic addition-elimination reactions, including that with hydrazine derivatives. (See (d), below.) It is important however to realise that where the >C=O group occurs in combination with other heteroatom bonds to carbon in functional groups such as –COOH (acids), - COOR′ (esters), -COCl (acid chlorides), -$CONH_2$ (amides), and so on, it does not behave in the same way and none of the above listed functional groups gives a derivative of this type.

(a) This has been dealt with in Topic 15.1 (e), but it is worth noting that in the sequence

$$CH_3CH_2OH \xrightarrow{[-2H]} CH_3CHO \xrightarrow{[+O]} CH_3COOH$$

the first oxidation step may be seen as a loss of hydrogen (dehydrogenation) and the second as an addition of oxygen. In fact the first step may be carried out industrially as a dehydrogenation over copper.

(b) The difference in ease of oxidation of aldehydes and ketones, referred to in Topic 15.1, results in only the former being oxidised by mild oxidising agents such as Ag^+ ($E^\ominus$ for $Ag^+ + e^- \longrightarrow Ag$ being 0.80V) and Cu(II) (which is reduced to Cu(I)). Thus Tollens' reagent (ammoniacal silver nitrate) on gentle warming is readily reduced from Ag^+, to Ag metal, giving the familiar 'silver mirror', with the aldehyde being oxidised to the corresponding carboxylic acid. Similarly aliphatic aldehydes will reduce Fehling's solution whose deep blue colour, due to Cu(II), is discharged, giving a red-brown precipitate due to copper(I) oxide, Cu_2O.

(c) Reduction may be carried out directly using H_2 and a Ni catalyst but the use of hydride reagents is more convenient. All of these – LiH, $LiAlH_4$, and $NaBH_4$ – essentially contain H^- which donates an electron and is oxidised to H_2 (oxidation state change $-I \rightarrow 0$). $NaBH_4$ and $LiAlH_4$ are preferred since they are more stable, and $NaBH_4$ has the advantage of being stable in the presence of water, whereas both LiH and $LiAlH_4$ react with it, as in $LiH + H_2O \rightarrow LiOH + H_2$. Thus if $LiAlH_4$ is used it must be in anhydrous solution (ethoxyethane) whereas $NaBH_4$ may be in aqueous methanol.
These reagents therefore effectively reverse the oxidations described in Topics 15.2(a) and 15.1(e) so that aldehydes thereby produce primary alcohols and ketones yield secondary alcohols. (See also reaction diagram on page 25, Topic 15.1 (b).)

(d) The reason for selecting such a complicated looking compound is that the reaction product with both aldehydes and ketones is readily obtained pure in a quantitative amount, and has a melting temperature characteristic of the particular carbonyl compound, which may thus be identified. The reaction, which is quite general in type, may be followed more simply using the parent molecule hydrazine itself (at least on paper, since hydrazine is a powerful rocket fuel!)

$$\underset{H}{\overset{CH_3}{>}}C=O + NH_2-NH_2 \longrightarrow \underset{H}{\overset{CH_3}{>}}C\underset{NHNH_2}{\overset{OH}{<}} \longrightarrow \underset{H}{\overset{CH_3}{>}}C=N-NH_2 + H_2O$$

The first step is addition across the C=O bond to form an intermediate which rapidly eliminates water to give the hydrazone. The <u>overall process is thus an addition – elimination, or condensation, reaction</u>. The 2,4-DNP reaction is identical except that a hydrogen atom on the hydrazine is replaced by $-C_6H_3(NO_2)_2$.

As non-syllabus background, the mechanism of the addition involves a nucleophilic attack by the nitrogen lone pair on the carbonyl carbon, which is δ+.

(e) Both aldehydes and ketones react with HCN to produce addition products according to

$$RCHO + HCN \rightarrow RCH(OH)CN$$

and

$$RCOR' + HCN \rightarrow RCR'(OH)CN$$

the reaction being initiated by attack of the CN⁻ ion at the carbon atom of the carbonyl group. The reaction is thus an example of <u>nucleophilic addition</u> and can be represented schematically as

$$\underset{R'}{\overset{R}{>}}\overset{\delta+}{C}=\overset{\delta-}{O} \quad \overset{NC-H}{\underset{}{|}} \quad \rightarrow \quad \underset{R'}{\overset{R}{>}}C\underset{OH}{\overset{CN}{<}}$$

The reaction products, sometimes known as cyanohydrins, are synthetically useful since they can be converted to <u>α-hydroxycarboxylic acids</u> by the usual hydrolysis of the nitrile group. (See Topic 16.2(a).)

(f) This is another of the interesting and distinctive tests associated with carbonyl compounds and often arises in organic problems, where some care is needed in deciding whether a particular compound will or will not give a positive test. This falls into two parts.

1. <u>A CH_3 group must be directly attached to the carbonyl</u>: this enables us to distinguish between various pairs of compounds such as

$$CH_3CH_2CH_2COCH_3 \text{ (yes) and } CH_3CH_2COCH_2CH_3 \text{ (no)}$$

Of the aldehydes, therefore, <u>only CH_3CHO gives a positive result</u>.

2. Alcohols containing a $CH_3CH(OH)$ group will be oxidised to contain a CH_3CO group by the reagent which contains positive iodine (+I) so that ethanol and secondary alcohols having OH in the 2-position will give positive results.

The distinctive yellow crystals and odour of the iodoform (CHI_3) give a clear test.

<u>Not</u> in the syllabus but a fascinating organic problem is the question of how one gets from CH_3CHO to CHI_3. Who could guess that result? The mechanism seems to be as follows:

$$R-\overset{O}{\underset{\|}{C}}-\underset{\underset{OH^\ominus}{\overset{H}{|}}}{CH_2} \rightleftharpoons R-\overset{O}{\underset{\|}{C}}-\overset{\ominus}{CH_2} \xrightarrow{I^+} R-\overset{O}{\underset{\|}{C}}-CH_2I$$

The two other hydrogens are similarly replaced → $RCOCI_3$. Then

$$R\overset{O}{\underset{\|}{C}}-CI_3 \xrightarrow{} R\overset{O}{\underset{\|}{C}}-OH + CI_3^\ominus \rightleftharpoons RCOO^\ominus + CHI_3$$
$$OH^\ominus$$

Topic 15.3 Carboxylic Acids and Their Derivatives

In this section are treated the properties of carboxylic acids and of certain of their (rather numerous) derivatives. These latter include acid chlorides, anhydrides, esters and amides and there are a number of unifying features as regards their reactions and relative reactivity which will be indicated below and especially in 15.3(b). It is recommended that the reactions of all these functional groups should be viewed in this light since such an approach may be more fruitful than viewing their reactions as individual and unconnected aspects of their behaviour.

(a)(i) Carboxylic acids contain the functional group –COOH, and those of general formula RCOOH, where R is an alkyl group, are generally much less volatile than aldehydes and haloalkanes of comparable relative molecular mass (e.g. ethanoic acid, CH_3COOH, has b.t. 118°C compared with propanal, CH_3CH_2CHO, b.t. 50°C and chloroethane, b.t. 12°C), due to the strong $O-H\cdots O$ hydrogen bonding in these compounds. Alkyl carboxylic acids are generally quite soluble in water, by virtue of strong water to carboxylic acid hydrogen bonding, but, except in this and other solvents in which similar strong solute–solvent interaction is possible, carboxylic acids usually exist largely as hydrogen bonded dimers, as shown below.

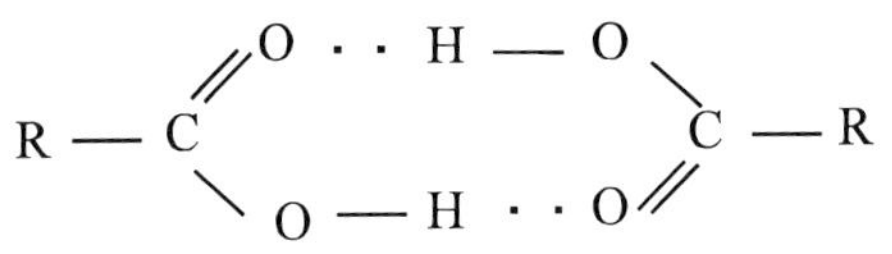

$$R-C\underset{O-H\cdots O}{\overset{O\cdots H-O}{\diagup}}C-R$$

The straight chain alkyl carboxylic acids are liquids for the lower members, but the boiling temperatures increase steadily as the chain lengthens and, for C_2 and above, are solids at room temperature. Moreover, as the chain length increases, the solubilities in water of the alkyl carboxylic acids also decrease: essentially the solubility conferred by the –COOH group, and its capacity to interact with the solvent by hydrogen bonding, becomes progressively less effective as the size of the inert hydrocarbon residue increases.

Note of course that similar considerations regarding volatility, solubility and acidity (see (ii), below) also apply to benzoic acid, C_6H_5COOH, and to other cognate carboxylic acids containing aryl groups.

(a)(ii) Students should here understand clearly the reasons why carboxylic acids are actually moderately strong acids (although not of course 'strong acids' in the sense that the term is defined in Topic 9 (q.v.)) and in particular should be able to recall and rationalise the relative acidities of carboxylic acids, phenols (as exemplified by phenol itself), alcohols (as exemplified by methanol) and water. Thus attention should first be directed towards the nature of the equilibrium $RCOOH + H_2O \rightleftharpoons RCOO^- + H_3O^+$ and why this dissociation should be (relatively) quite appreciably favoured.

Thus a superficial consideration of the structures of the carboxylic acid and its anion suggests that both contain a carbon-oxygen double bond, >C=O, and a carbon-oxygen single bond $-\overset{||}{C}-O$, as depicted below:

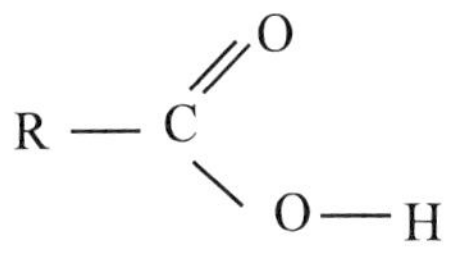

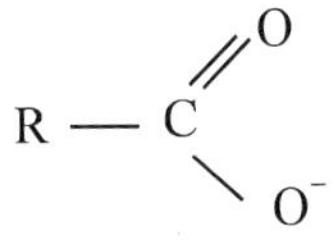

$$R-C\overset{\diagup O}{\diagdown O-H} \qquad R-C\overset{\diagup O}{\diagdown O^-}$$

However, actual measurements of such bond lengths (by X-ray diffraction and other techniques) shows that, as expected, in carboxylic acids the C=O double bond length at around 123 pm is substantially shorter than the C–O single bond length at about 133 pm, but that in the corresponding anion both carbon-oxygen bond lengths were the same, at some 128 pm, intermediate between the double and single bond lengths.

In fact, in the anion, the π-electron system is delocalised over the carbon atom and both oxygen atoms, which are clearly here equivalent and may be represented as shown below:

In this way therefore the anion is very substantially stabilised, thus facilitating the displacement of the equilibrium to the right hand side and leading to the appreciable acidity of the carboxylic acids. In terms of the nomenclature of Topic 9 the carboxylic acids are only 'weak acids' but ethanoic acid, with K_a = 1.85 × 10^{-5} mol dm^{-3} is fairly typical of such systems and carboxylic acids almost always have K_a values appreciably greater than that of carbonic acid, H_2CO_3, and thus readily liberate CO_2 from sodium hydrogencarbonate. [Electron supplying groups tend of course to decrease the acidity of the carboxylic acids so that K_a, which is 1.77 × 10^{-4} mol dm^{-3} for methanoic acid, decreases somewhat to the value given above for ethanoic acid when the electron supplying methyl group is substituted for the hydrogen attached to the carbon atom in HCOOH. Conversely, electron withdrawing groups such as the chloro group facilitate the loss of the proton, so that K_a for chloroethanoic acid, $ClCH_2COOH$, increases to 1.40×10^{-3}.]

Water of course is known to be only very weakly ionised, as reflected by the value of its ionic product, K_w = [H_3O^+][OH^-], which is only 1.0 × 10^{-14} mol^2 dm^{-6}, but methanol, which behaves like a neutral compound for all practical purposes, has in fact even less tendency to release a proton by virtue of the attachment of the electron releasing methyl group. Thus, alcohols in general are even weaker acids than water and their tendency to form alkoxide ions, of the form OR$^-$ for an alcohol ROH, is very small indeed when R is an alkyl group. However, when R is an aromatic residue, for example the C_6H_5 phenyl group, the situation is very different so that phenol, C_6H_5OH shows quite appreciable acidic properties.

Thus, when a hydroxy group is attached to the benzene ring the oxygen atom of the –OH group is more electronegative than carbon and thus attracts electrons more strongly. However, the oxygen atom also carries a lone pair of electrons which can be fed back into the delocalised π-electron system of the ring, and in this way eight electrons, six from the benzene ring system and the two of the lone pair, can be delocalised over seven atoms.

Thus, in phenol itself, this delocalisation leads to excess electron density over the benzene ring, thus facilitating electrophilic substitution (c.f. the facile reaction of phenol with bromine; Topic 15.1 (h)). Moreover, when considering the possible dissociation of phenol according to:

$$\text{C}_6\text{H}_5\text{OH} + \text{H}_2\text{O} \rightleftharpoons \text{H}_3\text{O}^+ + \text{C}_6\text{H}_5\text{O}^-$$

it is clear that the negative change on the phenoxide anion, $C_6H_5O^-$, can be dispersed and thereby stabilised by such delocalisation over the aromatic ring. In this way the anion is stabilised and the ionisation of phenol becomes relatively more favourable than for aliphatic alcohols, ROH. This stabilisation is appreciably less than that which occurs in the carboxylic anion, but is sufficient to make phenol markedly more acidic than both methanol and water, the relevant K_a values being 6.1×10^{-16} mol dm^{-3} for CH_3OH and 1.3×10^{-10} mol dm^{-3} for C_6H_5OH. Note however that phenol from this value is clearly less acidic than carbonic acid and so will not liberate CO_2 from sodium hydrogencarbonate. Students should therefore remember the sequence of decreasing K_a values for the series ethanoic acid > phenol > water > methanol and be able to explain this as described above. [They might also usefully be familiar with the representation of K_a values in terms of the function pK_a (= $-\log_{10} K_a$), noting that smaller K_a values are reflected by larger pK_a values.]

(a) (iii) As well as recalling the distinction between carboxylic acids and the great majority of phenols described in (a)(ii), above, students should also remember the use of the iron(III) chloride test for phenols described in Topic 15.1(i). They should also note that bi-functional species such as 2-hydroxybenzenecarboxylic acid (salicylic acid), $C_6H_4(OH)COOH$ and 2-hydroxybenzaldehyde (salicylaldehyde), $C_6H_4(OH)CHO$, will also give a positive test (purple colour) with this reagent for the present of a phenolic – OH group.

(b)(i) Students should recall that aldehydes are readily oxidised to carboxylic acids and that these acids are the eventual oxidation products of primary alcohols. They should be aware that species such as RCH_2OH are readily oxidised, first to aldehydes, RCHO, and eventually to carboxylic acids, RCOOH. They should recall that this complete oxidation is most readily brought about by heating under reflux with acidified dichromate(VI),

$Cr_2O_7^{2-}$, solution, but that other oxidants, such as acidified manganate(VII), MnO_4^-, may also be used. (Compare also Topic 15.1 (e).)

(b)(ii) The methyl side chains in aromatic compounds such as methylbenzene (toluene), $C_6H_5CH_3$, are normally fairly resistant to attack, apart from halogenation (c.f. Topics 8(a) and 14(a)). However, such groups can be converted to carboxylic acid groups by vigorous oxidation with manganate(VII) ion under alkaline conditions, followed by acidification to liberate the free acid from its salt. In the laboratory this is usually effected by lengthy refluxing with a mixture of potassium manganate(VII) and sodium carbonate, and subsequent acidification with hydrochloric acid. In this way, for example, methylbenzene can be converted into benzene carboxylic acid according to

$$C_6H_5CH_3 \xrightarrow{KMnO_4/Na_2CO_3} C_6H_5COONa \xrightarrow{H^+} C_6H_5COOH$$

(b)(iii) In considering the reactions listed under this heading it is useful first to examine the nature of the many processes which involve initial attack at the carbon atom of the carbonyl group. [Since the >C=O group is polarised in the sense $\overset{\delta+}{C} - \overset{\delta-}{O}$, many of these are initiated by the attack of a nucleophile at the carbon atom, for example the formation of 2,4-dinitrophenylhydrazones of aldehydes and ketones. (See Topic 15.2 (d).) Similarly, many of the reactions of carboxylic compounds begin with attack by a nucleophile at the carbonyl carbon atom, but, unlike the addition – elimination (condensation) reactions of aldehydes and ketones, are essentially nucleophilic substitution reactions, in which one group attached to the carbonyl carbon atom is replaced by another.

These reactions may be represented by the sequence

$$R-C\!\!\begin{array}{c}\diagup O\\ \diagdown O\end{array}$$
$$R-C\!\!\begin{array}{c}\diagup\\ \diagdown O\end{array}$$

where the attacking nucleophile, Y, may be either a neutral or a negatively charged species.]

In fact it is found that some groups attached to the >C=O group are more readily displaced than others, for example Cl is readily displaced as Cl⁻, so that in this case acid chlorides, RCOCl, are extremely reactive as a consequence. The reactivity series for the carboxylic acids and their derivatives is thus found to follow the sequence:

$$\text{RCOCl} > \text{RCO–O–COR} > \begin{array}{c}\text{RCOOH}\\ \text{RCOOR'}\end{array} > \text{RCONH}_2 > \text{RCOO}^-$$

acid chloride > acid anhydride > acid or ester > amide > carboxylate

The most reactive species thus lie to the left of the sequence and in general, although there are some exceptions, species to the right of the sequence are usually readily obtained from those to the left of them. It is suggested therefore that this unified approach may be of help to students in remembering and understanding some of the reactions of carboxylic acid derivatives covered in this Topic and in Topics 16.1 and 16.2 (below).

Students should therefore be able to recall methods of converting carboxylic acids (1) to esters and (2) to acid chlorides, together with the hydrolyses of these products and should be able to use such information in solving organic problems.

(1) The preparation of esters from carboxylic acids involves a procedure which is essentially the reverse of the hydrolysis reaction. Thus, in principle, a carboxylic acid, RCOOH and an alcohol, R'OH, may react together to form an ester, RCOOR', which conversely may be hydrolysed back to these same products, according to

$$\text{RCOOH} + \text{R'OH} \rightleftharpoons \text{RCOOR'} + \text{H}_2\text{O}$$

and normally such a reaction will reach an equilibrium situation somewhere between complete esterification and complete hydrolysis. The rate of attainment of equilibrium is however strongly acid catalysed and esterification is therefore favoured by the addition of a little concentrated sulphuric acid to the acid / alcohol mixture, which will also help to displace the equilibrium to the right by removing some of the water formed in the forward reaction. Another technique for esterification which essentially operates in the same way is known as the Fischer-Speier method in which the acid / alcohol mixture is saturated with anhydrous (gaseous) hydrogen chloride. In both methods an excess of the alcohol (which usually can readily be removed later by distillation) is often used, again to help drive the equilibrium to the right.

Conversely, although ester hydrolysis can be brought about by heating with excess aqueous acid, this is usually slow and hydrolysis by heating with aqueous alkali – hydroxide ion – is much more often used and quicker and more effective. Note that this is in accordance with the reactivity series given above since the carboxylic anion is the most stable of all the derivatives listed.

It should thus be noted that alkaline hydrolysis, for example <u>by sodium hydroxide,</u> will yield, from RCOOR', the alcohol, R'OH, and the sodium salt, RCOONa, of the carboxylic acid. Such sodium salts will be fairly soluble in water, even for aromatic species such as sodium benzoate, C_6H_5COONa. Since the carboxylic acids are (relatively) weak acids they may be displaced by treatment of such salts with a strong acid such as <u>dilute HCl</u> or H_2SO_4, but here aromatic acids such as benzoic acid are usually only sparingly soluble in water and will thus be precipitated on the acidification of solutions of their salts.

Thus, for example,

$$RCOOR' + NaOH \rightarrow RCOONa + R'OH$$

and $\quad RCOONa + HCl \rightarrow RCOOH + NaCl$

(2) As might be expected from the reactivity sequence given above, the preparation of acid chlorides from carboxylic acids requires the use of rather vigorous reagents. Thus the replacement of the hydroxyl group of a carboxylic acid by chlorine requires treatment, usually under reflux, with either phosphorus pentachloride, PCl_5, or sulphur dichloride oxide (common name thionyl chloride), $SOCl_2$, according to:

$$RCOOH + PCl_5 \rightarrow RCOCl + POCl_3 + HCl$$
$$RCOOH + SOCl_2 \rightarrow RCOCl + SO_2 + HCl$$

<u>The latter is actually somewhat the cleaner reaction since thionyl chloride is a readily volatile liquid and all the by-products are gaseous.</u>

Acid chlorides are very reactive acylating agents and (see again reactivity sequence) react readily with water, alcohols and ammonia to produce respectively carboxylic acids, esters and amides (see also Topic 16.2) according to:

$$RCOCl + H_2O \rightarrow RCOOH + HCl$$
$$RCOCl + R'OH \rightarrow RCOOR' + HCl$$
$$RCOCl + NH_3 \rightarrow RCONH_2 + HCl$$

They also react readily with amines (see Topic 16.1 (c)) and with carboxylate salts to yield anhydrides (see 15.3(c), below) according to:

$$RCOCl + RCOONa \rightarrow NaCl + RCOOCOR$$

(See also, again, the reactivity sequence.)

(b)(iv) In (b)(i), above and Topic 15.1 (e) it was seen that primary alcohols, RCH_2OH, could quite readily be oxidised to the corresponding carboxylic acids, RCOOH. Students should also recall that carboxylic acids may conversely be reduced to primary alcohols and be able to use and apply this knowledge to a variety of situations in organic chemistry. They should be aware that this transformation requires the use of the very powerful reducing agent, lithium tetrahydridoaluminate(III), $LiAlH_4$, and is not brought about by the milder reducing agent, sodium tetrahydridoborate(III), $NaBH_4$. It should also be noted that, whereas sodium tetrahydridoborate(III) is used in aqueous ethanolic solution (see Topic 15.2 (c)), lithium tetrahydridoaluminate(III) is such a strong reductant that it will react with water to liberate hydrogen and must therefore be used as a suspension in an inert solvent such as diethylether followed by the addition of water to liberate the required product. Note also that it is usually impossible to arrest this reduction at the intermediate aldehyde stage, so that RCOOH is converted directly to RCH_2OH.

Students should also have knowledge of the use of the decarboxylation of carboxylic acids, especially as an aid to structure determination. Thus generally carboxylic acids, RCOOH, may be decarboxylated to yield the corresponding alkane, RH. This may be effected by heating the sodium or calcium salts with excess solid sodium hydroxide or calcium oxide but is most conveniently achieved by heating the sodium salt with soda lime (calcium oxide slaked with sodium hydroxide solution). In this way for example sodium ethanoate yields methane: $CH_3COONa + NaOH \rightarrow CH_4 + Na_2CO_3$. Similarly benzoic acid, C_6H_5COOH, may be decarboxylated to benzene and phenylethanoic acid, $C_6H_5CH_2COOH$, to methylbenzene (toluene), $C_6H_5CH_3$. In this way therefore hydrocarbons with one fewer carbon atom than the carboxylic acid are formed. The yields obtained are usually not very good, but the main value of the technique is diagnostic in revealing the hydrocarbon fragment present in any given carboxylic acid.

A common error with this reaction is the belief that during decarboxylation carbon dioxide is evolved: it is not. The elements of CO_2 are indeed removed from the carboxylic acid, but the CO_2 is taken up by reaction with the alkali: it is the remaining hydrocarbon which is evolved.

(c) Students should be aware of the substantial use made in industry of ethanoic anhydride as an ethanoylating agent. Thus (see reactivity sequence in (b)(iii),above), in this context ethanoic anhydride is second only to ethanoyl chloride in reactivity and converts hydroxy groups in species of the form ROH to esters according to:

$$ROH + (CH_3CO)_2O \rightarrow ROCOCH_3 + CH_3COOH$$

(The use of ethanoyl chloride for this purpose would produce gaseous HCl which would be more difficult to cope with – $ROH + CH_3COCl \rightarrow CH_3COOR + HCl$.)

The main industrial use of ethanoic anhydride is in the conversion of cellulose to cellulose acetate for synthetic fabrics and smaller quantities are used for the preparation of vinyl acetate and aspirin. (Historical note: 'Acetate' is the older name for ethanoate, CH_3COO- and 'vinyl' denotes the group $CH_2=CH-$.) Note that anhydrides find quite general synthetic use in the conversion of hydroxy groups to esters, ROH → CH_3COOR (see Topic 17(j) for the preparation of aspirin) and also for the acylation of amino groups, $-NH_2$ → $-NHCOCH_3$ (c.f. Topic 16.1 (c)). On the whole, acid anhydrides are usually a somewhat less vigorous but often preferrable alternative to acid chlorides as acylating agents.

Students should also be aware of the great industrial importance of polyesters, of which terylene is a typical example. This compound was so named because it was first produced by the reaction between ethane-1,2-diol, $HOCH_2CH_2OH$, and benzene-1,4-dicarboxylic acid $HOOCC_6H_4COOH$, for which the trivial name is terephthalic acid. This reaction is an esterification reaction so that one molecule of benzene-1,4,-dicarboxylic acid could react with two molecules of ethane-1,2-diol according to

$$HOOC-C_6H_4-COOH + 2HOCH_2CH_2OH \longrightarrow$$

$$HOCH_2CH_2OOC-C_6H_4-COOCH_2CH_2OH$$

which could then further esterify via the free terminal –OH groups with further benzene-1,4-dicarboxylic acid molecules to give an infinite chain, linked by repeated ester groupings. This reaction is an example of a condensation polymerisation: see Topic 17(i).

Topic 16   Organic Compounds Containing Nitrogen

In Topic 16 the concern is with the organic chemistry of compounds containing nitrogen and in Topic 16.1 the interest is in species containing the amino groups, $-NH_2$, and with the base character of amines and other amino compounds in general. Here learning outcomes (a) – (d) deal with the properties of both aliphatic and aromatic amines and (d) is especially concerned with differences between them. By contrast, in (e) – (h) there are considered the unusual properties of amino acids, which contain both an acidic and a basic functional group, and the properties of peptides and proteins derived by the combination of two or more amino acids.

In Topic 16.2 the emphasis is on amides, which contain both a >C=O and an $-NH_2$ function, and on nitriles, which in many ways provide a synthetic pathway between simple haloalkanes and a range of carboxylic acid derivatives and other functions.

Topic 16.1   Primary Amines and Amino Acids

(a) Aliphatic amines, of general formula RNH_2, may most simply be regarded as derived from ammonia, NH_3, by the replacement of one hydrogen atom by the alkyl group, R. In fact the simplest approach to their preparation is the treatment of alkyl halides (haloalkanes) RCl, RBr or RI with ammonia, either aqueous or alcoholic, in sealed tubes at 100°C, according to:

$$RHal + NH_3 \rightarrow RN^+H_3Hal^-, \text{ followed by}$$
$$RN^+H_3Hal^- + H_2O \rightleftharpoons RNH_2 + H_3O^+ + Hal^-$$

However, although simple in principle, this method does not always give very good yields since more than one hydrogen of the ammonia is sometimes substituted by the alkyl group and it is important that the ammonia and not the alkyl halide should be in excess to counteract this. Amines can also be obtained by the reduction of nitriles [and of amides] with $LiAlH_4$ (see Topic 16.2(a)(ii)), in rather better yield, and by a variety of other indirect methods, but those are beyond the scope of this syllabus.

Candidates should note (see (b), below) that these amines are, like ammonia, basic: they are, like ammonia, rather weak bases, but because of the electron supplying action of the alkyl groups the aliphatic RNH_2 species are usually slightly stronger bases than ammonia. Note that, being basic, all amines readily form salts of the form $RNH_3^+X^-$ e.g. X = Cl for the addition of strong acids such as HCl.

Aromatic amines such as phenylamine (aniline), $C_6H_5NH_2$ are generally rather easier to prepare than their aliphatic counterparts since they can readily be obtained by reduction of the corresponding nitro compounds, e.g. $C_6H_5NO_2$, nitrobenzene, which can readily be prepared by the direct nitration of benzene. (See Topic 13(g)) Thus nitrobenzene, b.t. 211°C, is easily reduced by heating under reflux with granulated tin and concentrated hydrochloric acid. Students should note that, contrary to a popular fallacy, tin is not acting as a catalyst here: tin metal plus concentrated hydrochloric acid constitute a reducing agent and in fact tin(II) chloride in acid media can also be used to effect the reduction of many aromatic nitro compounds. Students should also appreciate that, in an acidic medium, the amine $C_6H_5NH_2$, is produced as the salt, $C_6H_5NH_3^+Cl^-$. In general such salts are readily soluble in water and to liberate the free amine the solution must be made alkaline, e.g. by the addition of excess sodium hydroxide. The liquid phenylamine (aniline), with b.t. 184°C is only very sparingly miscible with water and is then most readily recovered from the solution by steam distillation. This not only ensures that it may be distilled at only 100°C and then separated from water with a separating funnel, but avoids the need to distil the pure liquid at the higher temperature of 184°C and thereby minimises possible losses of product due to partial thermal decomposition.

(Note that steam distillation is a technique of general usefulness for high boiling liquids which are (i) essentially immiscible with water, and (ii) not decomposed by water.)

(b) Although nitrogen is more electronegative than carbon (see Module CH1, Topic 3.1(h)), it is much less so than oxygen, so that there is little tendency for heterolytic cleavage of a C–N bond to occur and amines do not undergo nucleophilic substitution reactions. In fact the lone pair on the nitrogen is less strongly held by the nucleus than the lone pairs of oxygen, so that a characteristic property of an amine is its tendency to act an electron pair donor and more specifically as a base.

Students should therefore recall that amines are weak bases, which generally give alkaline solutions and are partially ionised in water according to:

$$RNH_2 + H_2O \rightleftharpoons RNH_3^+ + OH^-$$

and, as noted in (a), above, they react with strong acids such as HCl, to give salts, such as $RNH_3^+Cl^-$, which are in fact equivalent to substituted ammonium salts.

As noted above all amines are weak bases (e.g. for $C_2H_5NH_2$ $K_b = 5.6 \times 10^{-4}$ mol dm^{-3}) but an aromatic amine, such as phenylamine, is markedly weaker than the aliphatic amines, so that, for $C_6H_5NH_2$, $K_a = 3.83 \times 10^{-10}$ mol dm^{-3}. This comes about because the lone pair on the nitrogen atom can be delocalised over the aromatic ring (c.f. phenol, Topic 15.3(a)(ii)) and thus is much less readily available to act as a proton acceptor, thereby reducing the base strength. [The base strengths of aromatic amines may however be markedly influenced by the presence of either electron withdrawing substituents (ii) (weaker) or electron supplying substituents (stronger) in the aromatic ring.] (iii)

(c) Students should note that the syllabus coverage of amines is restricted to primary amines of the form RNH_2 and that secondary and tertiary amines of the form RR'NH and RR'R"N are excluded. Thus all primary amines are readily acylated either by the most vigorous reagent, an acid chloride such as CH_3COCl (ethanoyl chloride) or by the slightly milder acid anhydrides (c.f. reactivity sequence, Topic 15.3 (b)(iii)). The reaction with ethanoyl chloride follows this course:

$$RNH_2 + CH_3COCl \rightarrow RNHCOCH_3 + HCl$$

the hydrogen atom attached to the nitrogen atom being replaced by the acyl (ethanoyl, CH_3CO-) group. In the laboratory considerable care is required in the use of ethanoyl chloride since this is a low boiling liquid, b.t. 52°C, and is extremely readily and violently hydrolysed by water to HCl and CH_3COOH.

[Note that acylation by acid chlorides yields HCl as a by-product which would then form a salt with another molecule of the amine so that, unless means were found to remove the

HCl produced, only one of two molecules of the amine would actually be acylated. For this reason acylation by acid anhydrides, e.g. ethanoyl anhydride, $(CH_3CO)_2O$, is often preferred since the molecule of ethanoic acid produced as the by-product does not consume a molecule of the amine in this way.

It may also be noted that acyl derivatives of amines are usually solids whereas the amines themselves are liquids, so that such derivatives are often used to characterise (by their m.t. values) the amines in question. This is becase amines themselves, RNH_2, are only relatively weakly hydrogen bonded by $>$N—H- - N$<$ hydrogen bonding whereas their acyl derivatives, RNHCOR', are more strongly hydrogen bonded by $>$N—H- -O$=$C$<$ hydrogen bonds (c.f. hydrogen bonding in peptides, Topic 16.1 (f) and (g), below).]

(d) Students should be able to recall and describe the reactions of both primary aliphatic and primary aromatic amines with cold nitrous acid (nitric (III) acid). They should be able to compare these two types of reaction and appreciate the importance of this distinction.

Thus in general aliphatic amines can be taken as reacting with nitrous acid, even in the cold, according to

✻ $RNH_2 + HNO_2 \rightarrow ROH + N_2 + H_2O.$ ✻

[When the alkyl group, R, is small, CH_3- or C_2H_5- the above equation is a good description of the reaction but for longer alkyl chains the process follows a very complex path with many rearranged products resulting and elimination reations also occurring, leading to alkene formation. As a preparative route from amines to alcohols therefore this reaction is certainly not recommened for C_3 species and above: see Q9, A2, 1988 and answer thereto (RND Publications). Nitrogen evolution is however effectively quantitative.]

With aromatic amines however the temperature at which the reaction is carried out is extremely important. Thus, at temperatures below 10°C (usually in the 0 –10°C range) a stable intermediate, known as a diazonium salt, is produced. Thus, for phenylamine, reacting with nitrous acid in the presence of hydrochloric acid the course of the reaction may be represented by:

$C_6H_5NH_2 + HNO_2 + HCl \rightarrow C_6H_5N_2^+Cl^- + 2H_2O$

where the diazonium cation takes the form:

$$C_6H_5-N^+\equiv N$$

Whereas aliphatic amines do form transient diazonium salts under similar conditions, they are exceedingly unstable and decompose rapidly, even at 0°C, as described above. Aromatic diazonium salts are however considerably more stable due to the delocalisation of the π-electrons of the nitrogen–nitrogen bond over the aromatic ring (c.f. similar considerations with regard to phenols and aromatic amines, Topics 15.3 (a)(ii) and 16.1 (b), above). Thus, as long as the reaction temperature is kept below 10°C, such diazonium salts are relatively stable species and in fact are exceptionally useful intermediates in synthetic organic chemistry. Many of these applications are outside the scope of the syllabus, but the replacement of the $-\overset{+}{N} \equiv N$, diazo–, group by hydroxyl is in fact the result of simply warming the diazonium salt in water. Thus, for phenyl diazonium chloride, $C_6H_5N_2^+Cl^-$, the result is:

$$C_6H_5N_2^+Cl^- + H_2O \rightarrow C_6H_5OH + N_2 + HCl$$

and phenol is produced in good yield. Thus, were the reaction of phenylamine with nitrous acid to be carried out at room temperature or above it would of course produce phenol directly and would appear to follow the same route as for the aliphatic amines, with nitrogen being evolved at once.

Students should be aware that nitrous acid, HNO_2, is itself a rather unstable species which breaks down on warming, and they should also know that when carrying out the preparation of diazonium salts – the diazotisation of an amine – nitrous acid is prepared in situ by the action of dilute hydrochloric acid on sodium nitrite, $NaNO_2$, either as a solid or in solution, with appropriate cooling.

Students should also be especially careful as regards the naming of the compound of formula HNO_2. The older name of nitrous acid is still in common and acceptable usage although the systematic nomenclature now uses nitric(III) acid. Similarly the older terminology for the salts names $NaNO_2$ as sodium nitrite, whilst the systematic form now uses sodium nitrate(III). Either naming system is here acceptable, but students are warned that if HNO_2 is named systematically the inclusion of the '(III)' of 'nitric(III)' is essential: an unqualified 'nitric acid' can only mean HNO_3, since this is the common name for the compound which systematically is nitric(V) acid. Similarly for the salts the '(III)' must be included in 'nitrate(III)' if a salt such as $NaNO_2$ (as opposed to $NaNO_3$ – nitrate(V)) is intended. Recent examination scripts have shown increasing evidence of either confusion or carelessness in this respect.

Amongst the useful reactions of diazonium salts one of the most important for students to recall and understand is the coupling reaction with phenols which these species undergo.

Thus benzene diazonium chloride, $C_6H_5N_2^+Cl^-$ readily couples with phenol in alkaline solution to produce 4-(phenylazo)phenol according to:

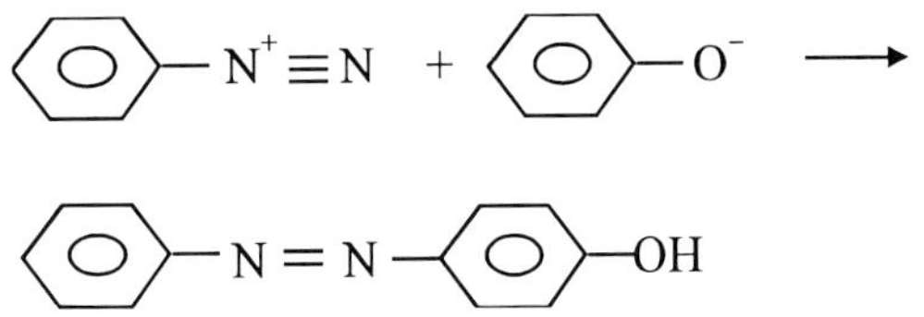

$$C_6H_5-N^+\equiv N + {}^-O-C_6H_4 \longrightarrow$$

$$C_6H_5-N=N-C_6H_4-OH$$

Note especially that the diazo group does not displace the –OH function, which is a common fallacy. The coupling usually takes place in the 4-position relative to the –OH substituent. In some systems, especially if the 4-position is already occupied or access to it hindered, coupling may take place at the position adjacent to the –OH group and in the naphthalene derivative, 2-naphthol, the coupling takes place in the 1-position, rather than in the unsubstituted ring. Thus:

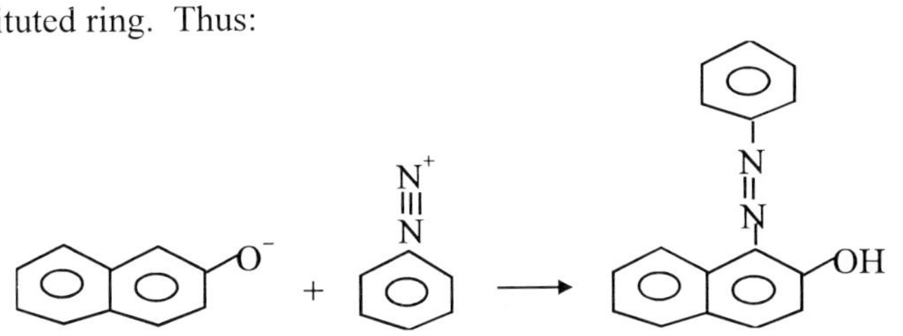

In general both phenols and aromatic amines may couple in this way with a diazonium salt but all the species so produced will contain two aromatic rings linked together by the –N=N– azo bridge, and in this way the π-electrons of the azo bridge may delocalise into either aromatic ring to yield an extended delocalised system.

As has been shown before in Topic 12(j), the –N=N– grouping is an effective chromophore and species such as azobenzene and the substituted azobenzenes produced by the coupling reaction are highly coloured, usually in the orange-red region (i.e. absorbing in the blue). Since the diazonium salt is formed from an aromatic amine, which clearly may contain other groups substituted into the ring, and it is to be coupled either with phenols or with other aromatic amines, which may equally well contain other substituents, it is clear that in this way a vast variety of variously substituted azobenzene (or similar aromatic) species may be prepared, and the colours of these compounds are appreciably affected by the presence of electron supplying or withdrawing substituents in the rings. By the latter half of the 19th century it had been found that these various azo species were both highly coloured and useful for dyeing fabrics, so that the coupling reaction is the essential process for the preparation of a great range of coloured species, now known collectively as azo dyestuffs.

Note that as well as being of great commercial importance the coupling reaction is extremely useful for the detection of aromatic amines, since such a compound may be diazotised and coupled with an aromatic hydroxy compound, such as 2-naphthol, to give a highly coloured, usually red, azo compound. On the other hand, aliphatic amines do not yield a stable diazonium compound and so do not undergo the coupling reaction nor give a highly coloured azo compound.

Preamble to (e) – (h)

Students should first of all appreciate that the subject matter of outcomes (e) – (h) concerns a very closely related group of substances – proteins, peptides and amino acids. Basically proteins are the essential constituents of all living matter. They constitute the major components of skin, hair, horn, muscle, haemoglobin, enzymes, viruses, blood and numerous other organisms. Essentially proteins are very long chain molecules consisting of numerous carbon atom chains of quite modest length linked together at many points by the –NH.CO– grouping. This is effectively the amide functional group (–$CONH_2$), dealt with in Topic 16.2, but in this present context the –NH.CO– link is usually known as a peptide link. Many of the properties of proteins and polypeptides are associated with this –NH.CO– linkage, including the way in which they behave on hydrolysis. Thus, the –NH.CO– linkage tends to break to yield both an amino, –NH_2, and a carboxylic acid, –COOH, function, so that if one represents a section of a protein as a series of chains connected by peptide links:

$\sim\sim\sim\sim$ NH.CO $\sim\sim\sim\sim\sim\sim\sim$ NH.CO $\sim\sim\sim\sim$ etc.

hydrolysis will yield

$\sim\sim\sim\sim$ NH_2 HOOC $\sim\sim\sim\sim\sim\sim\sim$ NH_2 HOOC $\sim\sim\sim\sim$ etc.

and if the process is complete the protein will be totally broken down into species containing both amino groups and carboxylic acid groups in the same molecule. These are known as amino acids and almost all of the systems of biochemical interest are what are known as α-amino acids in which both the amino group and the carboxylic acid group are attached to the same carbon atom. Most of the species of interest can thus be formulated as R–CH(NH_2)COOH in which the group R may be just hydrogen, a methyl group or a longer or branched carbon chain and in some cases additional functional groups may be present in the group R. A few of the more important α-amino acids are listed below as background information.

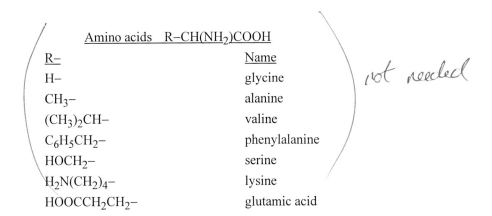

The names listed above are of course the trivial names for these compounds: they can be named systematically e.g. glycine is aminoethanoic acid, but these trivial names are almost universally used in the literature relating to these systems. Note that students are expected to <u>know</u> what an amino acid <u>is</u> and to be able to give the formulae for examples of amino acids. They are <u>not</u> however expected to know the trivial names for these amino acids nor to write down formulae corresponding to these trivial names.

<u>Peptides</u> may be regarded either as arising from the <u>partial hydrolysis of proteins,</u> with some –NH.CO– <u>linkages broken but others not,</u> or as being <u>produced by the combination</u> (with the elimination of water) <u>of two or more amino acids.</u> (The actual methods whereby amino acids may be so combined are unfortunately rather tedious and much synthetic effort, beyond the scope of this syllabus, has been expended on it.) Where more than two amino acids are so combined the products are known as <u>polypeptides</u> and the distinction between polypeptides, with, say, 50 or 100 amino acid residues, and proteins becomes increasingly blurred and artificial. Students should therefore be clear about the relationship between amino acids and peptides (see (f), below) and between peptides (and polypeptides) and proteins, and be aware of the biochemical importance of these species. Students should also note that compounds such as <u>nylon-6,6</u> (see Topic 16.2 (b)) and similar synthetic polyamide based fabrics also contain the –NH.CO– linkage and thus share certain properties of natural fabrics, especially the occurrence of $>\!\!N\!\!-\!\!H\cdot\cdot O\!=\!\!C\!<$ hydrogen bonding between the chains.

(e) As noted in the preamble above, the amino acids of interest are <u>α-amino acids,</u> of general formula RCH(NH$_2$)COOH (see Table, above) in which both the <u>amino and carboxylic acid functions are attached to the same carbon atom.</u> However, although these systems are usually represented as if they contained the basic –NH$_2$ and the acidic –COOH groups, they exist predominantly as internal salts known as 'zwitterions' which can be formulated as RCHNH$_3^+$COO$^-$. Moreover, to some extent, amino acids partake of amphoteric character in that in acid solution both functional groups are mainly protonated (as –NH$_3^+$ and –COOH) whilst in alkaline solution the reverse situation obtains (with –NH$_2$ and –COO$^-$).

[For all amino acids there will be an intermediate pH value at which the number of positively and negatively charged groups will be equal – not necessarily at pH7 – known as the isoelectric point, whose value depends on the K_a and K_b values for the –COOH and –NH$_2$ functions respectively. Candidates will not however be expected to carry out such calculations.]

Because of their two functional groups which may both hydrogen bond to water molecules, amino acids are generally easily soluble in water. However, because of their zwitterionic structure they are much less soluble in organic solvents and tend rather to resemble inorganic salts in this respect. This feature is also responsible for the fact that they mostly have high melting temperatures (e.g. glycine 232°C, alanine 295°C, valine 315°C) and often melt with decomposition.

(f) The simplest of all peptides are of course dipeptides in which two amino acids join together (with the loss of a molecule of water) to form a peptide link, –NH.CO–. Thus for glycine H$_2$NCH$_2$COOH one can visualise two such molecules combining to give the molecule H$_2$NCH$_2$CO.NHCH$_2$COOH but dipeptides can of course also be formed by the combination of two different amino acids. Thus, for example, a molecule of glycine, H$_2$NCH$_2$COOH, and a molecule of alanine, CH$_3$CH(NH$_2$)COOH, could combine to produce either CH$_3$CH(NH$_2$)CO.NHCH$_2$COOH as one dipeptide or H$_2$NCH$_2$CO.NHCH(CH$_3$)COOH as another. Both of these dipeptides of course have a free amino, –NH$_2$, function and a free carboxylic acid, –COOH, function and could therefore enter into combination with further amino acids to form tri- and higher polypeptides. Students should therefore be able to formulate the possible dipeptides which can be formed from any two different amino acids and also appreciate the way in which higher polypeptides are formed.

(g) Students should understand the manner, described in (f) above, in which polypeptides are built up, and should also appreciate that the boundary line between polypeptides and proteins is somewhat indistinct. However, as a rough rule, the smaller proteins usually have polypeptide chains with somewhere between 50 and 100 amino acid residues, whilst the larger proteins can contain several thousand such units.

The characteristic properties of any given protein depend upon the detail of its three-dimensional structure, which is usually described on three levels. Thus the primary structure of a protein is simply dependent upon the precise sequence of amino acids in the chain whilst the secondary structure describes the actual configuration of the chain: thus, for example, both zig-zag structures and coils (the now familiar helix) may be encountered in various different proteins. Various twists and convolutions of the chains may also be present and these may be either regular or irregular and are said to represent the tertiary structure of the system. This usually involves N–H--O hydrogen bonding which can occur either between the >N–H of the –NH.CO– peptide link of one chain and the >C=O of the –NH.CO– peptide link of another chain or between one –NH.CO– group of a chain and another –NH.CO– group of the same chain. In both cases the hydrogen bonding is between >NH and O=C< and is usually quite strong and important for determining the actual shape of the protein.

Schematic representations of the two types are shown below:

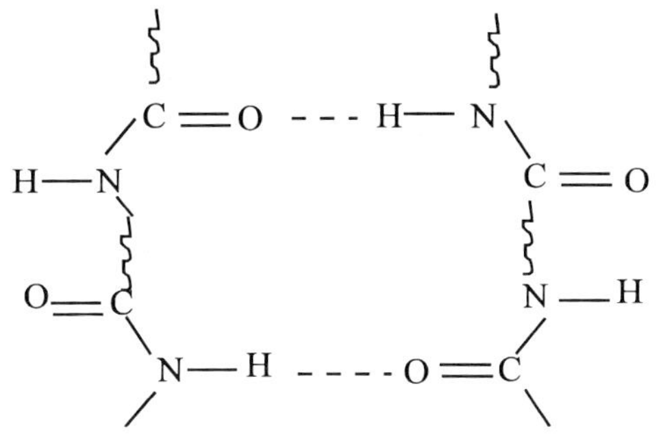

Hydrogen bonding between two different chains.

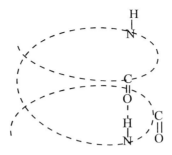

Hydrogen bonding between sites in the same chain.

[In addition to the common >NH--O=C< hydrogen bonding either within or between chains there are two other ways in which interactions between chains may occur. Thus amino acids such as lysine and glutamic acid have respectively additional $-NH_2$ and $-COOH$ groups which may also be involved in hydrogen bonding whilst the amino acid serine has a $-CH_2OH$ function which again may be involved in further hydrogen bonding. A more interesting possibility also arises when the amino acid cysteine, with the $-CH_2SH$ group is involved: in this situation two $-CH_2SH$ groups from different chains or from different sites within the same chain may undergo an oxidative coupling to form a disulphide bridge:

$$>CHSH \quad HSCH< \longrightarrow \quad >CH-S-S-CH<$$

thus forming an additional chemical bond between the sites.]

Finally it may be noted that the lower peptides are, like amino acids, usually fairly soluble in water, but less so in organic solvents. Moreover, this solubility decreases substantially as the chain is lengthened, but most tend, like amino acids themselves, to melt with decomposition somewhere between 200° and 300°C. The solubility and thermal properties of proteins are so diverse that no useful generalisation is possible.

(h) Students should have a general awareness of the importance of proteins in living systems, as outlined in the preamble to (e) – (h), above, and should know about some of the systems mentioned there. For example, fibres such as silk and wool are both protein fibres and these two substances also exemplify respectively the two types of hydrogen bonding (between chains and within chains) discussed above in (g) under tertiary structure.

In addition to this general background however students should also be familiar with another area involving protein chemistry, namely the nature and function of the substances, occurring in nature, which accelerate most of the chemical reactions which take place in the living cell. Effectively they are biocatalysts and are the means by which the living cell is enabled to carry out a vast range of complex chemical reactions and transformations. In general they facilitate reactions which occur at body temperature in the aqueous phase and at or close to neutral pH, and each enzyme is usually highly specific to only a small number of such reactions, and is usually quite sensitive to changes in temperature, pH and ionic environment.

One of the earliest studies of what is now called enzyme chemistry was concerned with the fermentation processes brought about by yeast and it was found that a substance could be isolated from the yeast cell which by itself could ferment sugar to give ethanol. Further biochemical transformations were then found also to be capable of being brought about by other biocatalysts and eventually such substances were found to be protein in nature.

It was soon found that such enzymes could effect reactions in living systems, under very mild conditions, which often required extremes of conditions (temperature, pressure) to bring about in the laboratory. Each enzyme only catalyses one or a very few reactions and the key to their activity is in their precise shape by which they can bond to the reacting species such that these molecules are exactly aligned for reaction to occur. Note also that the loss of activity of enzymes resulting from even quite gentle heating is consistent with this picture since even relatively small changes in temperature may disrupt the critical geometry of the enzyme molecule. Similarly, since proteins usually contain numerous amino and carboxylic acid functions, these will clearly be influenced by changes in pH, which may therefore influence the ability of the enzyme to catalyse the reaction in question.

Topic 16.2 Amides and Nitriles

This topic concerns two groups of compounds – amides and nitriles – and a short introduction is here appropriate. First of all students must avoid the all too prevalent confusion between amines and amides: an amine contains the functional group $-NH_2$ and an amide contains the functional group $-CONH_2$. There is probably no easy way to remember this distinction, but the recollection that amides are carboxylic acid derivatives whilst amines are not (see Topic 15.3) may help.

Another frequent cause of confusion concerns the base properties of the above two functional groups. Whilst all amines are to a greater (aliphatic) or lesser (aromatic) extent basic, amides show no significant tendency to act as bases. Thus, in the amide, $-CONH_2$, group the electron withdrawing tendency of the carbonyl, $>C=O$, group inhibits the availability of the nitrogen lone pair for protonation, so that any basicity of amides is negligible.

Nitriles constitute an important functional group for organic synthesis. Thus, as seen in Topic 14 (b), the halogen atoms in haloalkanes and related species may readily undergo nucleophilic subsitution by CN^- to form nitriles, and nitriles, of the form RCN, provide a number of routes by which an additional carbon atom can be added to the carbon chain in the alkyl group, R, and these methods are further discussed below.

(a) Students should be able to recall the processes listed in (i) and (ii) below and should be able to apply them to elucidate organic problems e.g. those involving a sequence of steps either to build up a molecule or to break it down into other simpler species, and should similarly be able to use this knowledge to deduce the identities of organic compounds from data about their reactions.

(a)(i) Students should be familiar with and able to describe two methods of converting carboxylic acids to amides. These are (1) the treatment of carboxylic acids with either $SOCl_2$ or PCl_5 to produce the corresponding acid chloride, i.e. RCOOH → RCOCl (see Topic 15.3 (b)(iii)), followed by the treatment of this acid chloride with concentrated ammonia, according to: $RCOCl + NH_3 \rightarrow RCONH_2 + HCl$

This reaction is usually fairly rapid and will often take place without warming. It is advisable however here to use concentrated (0.880) ammonia rather than the dilute solution since ammonia and water are both nucleophiles competing to attack the acid chloride and too much water would favour simple hydrolysis of the acid chloride to the carboxylic acid, rather than attack of NH_3 to give the amide.

Alternatively: (2) the carboxylic acid may be converted to its ammonium salt, $RCOONH_4$, by treatment with ammonia or by $(NH_4)_2CO_3$. The solid ammonium salt then, on strong heating over a period, will lose water according to:

$$RCOONH_4 \rightarrow H_2O + RCONH_2$$

affording an alternative method of preparation of an amide.

(a)(ii) As noted above the preparation of nitriles affords a method of adding an extra carbon atom to an alkyl chain. Thus for example a nitrile, RCN, obtained by the action of KCN on RBr, may be reduced on treatment with the powerful reducing agent $LiAlH_4$, in ethereal suspension, to an amine of formula, RCH_2NH_2. (The initial treatment with $LiAlH_4$ in ether then requires the addition of water to destroy the complex formed and liberate the free amine: note that water must be excluded from the reduction step since otherwise it would be reduced to hydrogen by the LiAlH$_4$.) [Amides, RCONH$_2$, may similarly be reduced to RCH$_2$NH$_2$ by LiAlH$_4$.]

Students should also recall that nitriles may additionally be hydrolysed to carboxylic acids by acid hydrolysis (warming usually needed) according to:

$$RCN + H_2O \rightarrow RCONH_2$$
$$RCONH_2 + H_2O \rightarrow RCOOH + NH_3$$

The reaction cannot usually be stopped at the intermediate amide step and of course, in the presence of acid, ammonia is here produced as an ammonium salt. Nitriles also readily undergo base hydrolysis, in a similar fashion, except that here, e.g. with NaOH, the product is the carboxylate salt, e.g. RCOONa, from which the acid itself may be liberated on the addition of a strong acid such as HCl.

Students should also be fully familiar with the hydrolysis of amides, which can also be carried out either with acid or base. Here however base hydrolysis is rather easier:

$$RCONH_2 + NaOH \rightarrow RCOONa + NH_3$$

and the evolution of ammonia on base hydrolysis is strongly indicative of the presence of an amide.

Thus, in summary, the hydrolyses of nitriles and amides follow the pattern:

	Base Hydrolysis (NaOH)	Acid Hydrolysis (HCl)
Nitriles (RCN)	RCOONa + NH$_3$	RCOOH + NH$_4$Cl
Amides (RCONH$_2$)	RCOONa + NH$_3$	RCOOH + NH$_4$Cl

(As noted above amides constitute intermediates in the hydrolysis of nitriles but can very rarely be isolated:

$$RCN + H_2O \rightarrow RCONH_2 \quad RCONH_2 + H_2O \rightarrow RCOOH + NH_3 \;)$$

(b) The properties of polyamides, containing the –NHCO– link were noted when dealing with peptides and related compounds with the same grouping (see Topic 16.1 (g) and preamble to (e) – (h)). Students should have some awareness of the mode of synthesis of such compounds and of their properties and industrial importance. Thus in the synthesis of one of the most important polyamides, nylon-6,6, the essential starting material is hexane-1,6-dioic acid, formulated as HOOC(CH$_2$)$_4$COOH. From some of this two synthetic steps yield 1,6-diaminohexane, H$_2$N(CH$_2$)$_6$NH$_2$, via the nitrile NC(CH$_2$)$_4$CN and subsequent reduction to H$_2$NCH$_2$(CH$_2$)$_4$CH$_2$NH$_2$ ≡ H$_2$N(CH$_2$)$_6$NH$_2$.

Hexane-1,6-dioic acid is then reacted with 1,6-diaminohexane in a condensation polymerisation (see Topic 17) in which a molecule of water is eliminated between each $-NH_2$ and $-COOH$ function according to:

$$nHOOC(CH_2)_4COOH + nH_2N(CH_2)_6NH_2$$
$$\downarrow$$
$$\text{\textthreesuperior}(CO(CH_2)_4CO.NH(CH_2)_6-NH\text{\textthreesuperior})_n$$

giving polymer chains of alternately four and six $-CH_2-$ groups, linked together by $-NHCO-$ linkages. Such synthetic polyamides possess many of the properties of natural fibres. Thus nylon forms a long chain molecule of considerable elasticity which can be spun into threads, but can also be moulded to form cogs and gears. It is also widely used in clothing (tights and stockings), in carpets and for climbing ropes.

Topic 17   Organic Synthesis and Analysis

A number of the subjects incuded in this Topic will already have formed the basis of questions previously set for the former WJEC A level Chemistry examinations, and examples embodying outcomes (a), (b), (f) and (g) may be sought therein. (See the series of Model Answers from 1988 onwards.) Outcome (d) is new to this syllabus but for outcome (j) only the outline incorporated into the present syllabus on p.37 is now required. As regards outcome (e), it has of course always been necessary for students to understand the purpose of the various operations of practical chemistry, and all that has now been done is to make these requirements rather more detailed and explicit.

(a)   Students should understand and appreciate the relationship between empirical formulae and elemental composition data, and should be able to derive empirical formulae from supplied elemental composition values. Usually, for organic compounds, this will be two or more elements from carbon, hydrogen, oxygen, nitrogen and the halogens (c.f. Topic 7(e)).

Students should be aware of the distinction between empirical formulae and molecular formulae. As well as being able to use the following types of data to obtain relative molecular mass values (c.f. Topic 2), candidates should also be able to use empirical formulae in conjunction with titration values, gas volumes, mass spectrometric ion values and gravimetric data in order to deduce molecular formulae. Examples of such data would include titration results for organic acids against sodium hydroxide or organic bases against hydrochloric acid (recognising their respective weak acid vs. strong base and strong acid vs. weak base character), the evolution of CO_2 from $NaHCO_3$ by organic acids or the evolution of N_2 by the action of HNO_2 on amines, m/e molecular and fragmentation patterns, and the masses of silver halides produced by solvolysis and subsequent treatment with silver nitrate of defined amounts of organic halo compounds. (Note that this list is intended to be exemplary, not exhaustive.)

(b) Students should be able to use mass spectrometric data to deduce the structures of simple organic molecules. Such data may include either or both of the molecular ions and the fragmentation patterns. Such molecules will exclude any cyclic structures and will contain no more than five carbon atoms. The species concerned may contain any number of the elements carbon, hydrogen, oxygen and nitrogen and may, in addition, also contain one atom of chlorine per molecule,(c.f. Topic 2).

It is useful to be able to recognise likely fragments rapidly, either from their m/e value directly or from the difference between an m/e value and that of the parent (molecular) ion. For example a 15 peak is usually CH_3 (although it could be NH if nitrogen were present), and a 43 peak in a spectrum whose molecular ion was 58 would show that a (58–43), i.e. 15, fragment had broken off, again identifying methyl. Common fragments include m/e/ 1 (H), 15 (CH_3), 17 (OH), 28 (CO), 29 (C_2H_5), 35 and 37 (Cl), 43 (C_3H_7), 44 (CO_2) and 45 (COOH), although these are not always unique to the species given, e.g. CH_3CO also has m/e 43. Chlorine fragments and chlorine-containing fragments are especially useful since there will always be two peaks differing by 2 units and in an abundance ratio of 3 : 1. This feature will of course show up in the molecular ion as well as in any chlorine-containing fragment such as CH_2Cl, which will give 49 and 51 peaks in the $^{35}Cl : ^{37}Cl$ ratio of 3 : 1.

Note that only singly charged m/e peaks will ever be given, doubly charged peaks are in any case always very small.

(c) Students will be required to interpret simple infrared spectra by the use of characteristic group frequencies. Such spectra will relate to the 1,500 to 4,000 cm^{-1} region and may be supplied either in numerical or diagramatic form. The characteristic group frequencies listed in the syllabus will be supplied within the appropriate examination paper. Students will not therefore be required to commit to memory the values of the wave numbers at which the relevant peaks occur, although it will obviously be advantageous if they are so familiar with them. Students should be aware that the interpretation of such infrared spectral data may well form part of a wider organic chemistry problem and should be prepared to use such information in conjunction with other information, both qualitative and quantitative. (See also Topic 12(k).)

(d) Since the n.m.r. technique was developed in the 1950s chemists (especially organic chemists) have benefited enormously. This is primarily because of the vast amount of information about the nature, number and location of the constituent hydrogen atoms which can be gleaned from the ^{1}H n.m.r. spectra of organic molecules.

The n.m.r. data thus yield information at three levels. In the first place the chemical shift (δ) value for a particular resonance may prove very helpful in deducing the environment of the atom (or atoms) responsible. Thus a great deal of data has been built up to indicate the likely values for particular groups or combinations of groups and some of these are shown in the Table below. Thus, for example, most aromatic protons lead to δ values in the 6 – 8.5 region whilst for aldehydes the peaks are usually in the 9 – 10 region and for carboxylic acids in the 10 – 12 region. In this way possible environments for the ^{1}H atoms can be suggested in much the same way as i.r. frequencies suggest the possible vibrational modes involved there.

At the second level, the intensity of any particular resonance, measured as the area under the curve for any given peak (or group of peaks), is directly proportional to the number of ^{1}H atoms involved. Thus, whilst benzene, C_6H_6, yields only a single n.m.r. peak at δ =7.3 (since all the ^{1}H atoms are equivalent), both methylbenzene (toluene) and 1,4 – dimethylbenzene (p-xylene) yield two, one typical of the ring protons and the other corresponding to those of the –CH$_3$ group(s). For methylbenzene the peak areas for δ = 7.2 and δ = 2.4 respectively are in the ratio 5 : 3, showing five ring protons to three methyl group protons, whilst for 1, 4-dimethylbenzene the δ values are 7.0 and 2.3 respectively in the ratio 2 : 3 (or 4 : 6), representing four ring protons to six methyl group protons.

There is however further information to be gathered at a third level, especially when high resolution instruments are used. Thus when non-equivalent nuclei (protons) are attached to <u>adjacent</u> atoms (such as C, N or O) in a given molecule, the magnetic moment of one may couple with that of the other, leading to the splitting of both peaks into multiplets, the multiplicity of each depending on the numbers involved. As examples the simple molecules of ethanol and ethanal are now treated.

The three peaks in the ethanol spectrum (centred around δ = 4.7, 3.6 and 1.15 respectively) show an intensity ratio of 1 : 2 : 3 and thus clearly correspond to the –OH, -CH$_2$- and –CH$_3$ contributions respectively. These peaks exhibit multiplicities (multiplet structure) of 3, 4 and 3 respectively and thus follow the general rule that indicates that a splitting into *n* components denotes the presence of *n*-1 ^{1}H atoms on the adjacent atom. Thus the two –CH$_2$ protons split the –OH peak into 3 and the three –CH$_3$ protons split the –CH$_2$ peak into 4. (The one –OH proton does induce a further splitting of the –CH$_2$ peak but this is only seen under high resolution.)

By comparison the n.m.r. spectrum of ethanal is very simple. The two groups of peaks at around δ = 9.5 and δ = 1.0 show an intensity ratio of 1 : 3 and so are clearly attributable to the –CHO proton and the –CH$_3$ methyl protons respectively. The δ = 9.5 peak is split into 4 by the –CH$_3$ protons and the δ = 1.0 peak into 2 by the –CHO proton. These results therefore underline the finding that the splitting of any resonance into *n* components indicates the presence of *n* – 1 hydrogen atoms on the adjacent carbon, nitrogen or oxygen atoms. Note that students are not expected to be able to derive this rule, only to apply it, nor are they expected to understand the subtle distinction between chemical and magnetic equivalence.

(See also Q.4(b) of Specimen Paper CH4, W.J.E.C., 2000, for a further example.)

Chemical Shifts of Common ^{1}H Groupings

Grouping	δ
-CH$_3$	0.9
-CH$_2$-	1.3
R$_3$C-H	1.5
Ar-CH$_2$-	2.3 – 3.0
Ar-H	6.0 – 8.5
RCH$_2$NH$_2$	2.0 – 2.8
R-NH$_2$	1.0 – 5.0
R-CH$_2$-CO-R'	2.0 – 2.7
RCH$_2$-OH	3.4 – 4.0
R-OH	1.0 – 5.0
R-CO-OCH$_2$-R'	3.7 – 4.1
RCH$_2$-COOR'	2.0 – 2.2
R-CHO	9.0 – 10.0
R-COOH	10.0 – 12.0

Note: the δ values listed represent only the commonly observed ranges: the limits are not to be taken as immutable.

(e) Students should be aware of the great desirability of integrating their experience of practical chemistry with their knowledge and understanding of the underlying theory. They must always resist the temptation to abandon one or other of these two aspects of the same subject at the laboratory door. They will therefore be expected to understand the reasons behind the choice of particular conditions for particular reactions and to appreciate the essentials of the techniques of manipulation, separation and purification used in organic chemistry, the most important of these being filtration, distillation and recrystallisation, together with the mechanical (tap-funnel) separation of one liquid from another with which it is immiscible.

A few examples of the sort of understanding needed are given below.

(1) The $Cr_2O_7^{2-}/H^+$ oxidation of primary aliphatic alcohols. Here the initial oxidation product from RCH_2OH is the aldehyde, RCHO. If the reaction product is at once distilled out of the reaction mixture, the fairly volatile aldehyde, RCHO (no hydrogen bonding) will be recovered, but if the reaction is allowed to continue under reflux further oxidation to the carboxylic acid, RCOOH, will ensue.

(2) In the absence of either acid or base the hydrolysis of an ester, RCOOR', will not go to completion but will reach an equilibrium according to:

$$RCOOR' + H_2O \rightleftharpoons RCOOH + R'OH$$

The hydrolysis usually proceeds most readily in the presence of a base such as NaOH, but the product is then not the acid, RCOOH, but its sodium salt, RCOONa. Such sodium salts are usually water soluble but, although many aliphatic acids are also water soluble, most aromatic acids are only sparingly so. Thus the base hydrolysis of an aromatic ester such as $C_6H_5COOC_2H_5$ produces a homogeneous solution containing C_6H_5COONa and C_2H_5OH, but on acidification (e.g. with HCl) the sparingly soluble benzoic acid, C_6H_5COOH, is obtained as a white precipitate. Since the solubility of benzoic acid increases appreciably with temperature, this precipitation is best effected with ice-cold solutions to ensure maximum recovery of the carboxylic acid.

(3) Amines are sometimes produced by reaction sequences beginning under acidic conditions. For example nitrobenzene, $C_6H_5NO_2$, is readily reduced by tin and concentrated hydrochloric acid. Here however, under acidic conditions, the initial product is not $C_6H_5NH_2$, but its salt, $C_6H_5NH_3^+Cl^-$, and the free base is only liberated on the addition of excess sodium hydroxide, according to:

$$C_6H_5NH_3^+Cl^- + NaOH \rightarrow NaCl + H_2O + C_6H_5NH_2.$$

(4) Confusion often arises over the nature of other hydrolyses carried out in the presence of acids or bases. Thus amides can be hydrolysed by either, although their base hydrolysis is both rather easier and readily diagnostic for the detection of an amide. Thus in the presence of an acid:

$$RCONH_2 + HCl + H_2O \rightarrow RCOOH + NH_4Cl$$

and in the presence of a base:

$$RCONH_2 + NaOH + H_2O \rightarrow RCOONa + NH_3 + H_2O$$

Clearly the detection of ammonia in the latter is a strong indication of the presence of an amide.

However, some years ago (1991, A1, Q9) a problem was set in which an amide, CH_3CONH_2, could be subjected to either base or acid hydrolysis; a depressingly large number of candidates chose the latter and asserted that ammonia would thereby be evolved! Note therefore that acid hydrolysis will not lead to the evolution of a basic gas, nor conversely will base hydrolysis result in the evolution of an acidic gas.

(5) Temperature control. Quite often this is important either for the yield or the purity of the product obtained. Thus for the diazotisation of aromatic amines by HNO_2, it is important that the reaction temperature is kept below 10°C: otherwise the diazonium salt, $ArN_2^+Cl^-$, will begin to decompose, mostly to give the phenol, for example

$$C_6H_5N_2^+Cl^- + H_2O \rightarrow C_6H_5OH + N_2 + HCl.$$

Similarly, for the nitration of benzene by the HNO_3/H_2SO_4 mixture, the temperature should not be allowed to rise above about 60°C or dinitration of the aromatic ring will begin to occur, giving 1,3-dinitrobenzene, $C_6H_4(NO_2)_2$, as well as nitrobenzene, $C_6H_5NO_2$.

At A/AS level candidates will usually have most experience with organic solids, rather than liquids, since the former are generally much easier to manipulate. Students should therefore be aware of the considerations regarding recrystallisation of such materials. Essentially potential impurities should be either insoluble or very soluble in the solvent chosen, so that they may either be removed by filtration or left behind in solution after recrystallisation. In addition the material to be recrystallised should have a large temperature dependence of solubility in that solvent, so that after being dissolved at a high temperature it may crystallise on cooling, leaving impurities in solution. Students should appreciate that solid derivatives are often prepared from liquids (e.g. 2,4-dinitrophenylhydrazones from aldehydes or ketones) so that the characteristic melting temperatures of these solids may enable the identities of the liquids to be established by comparison with literature values. For this purpose pure derivatives are essential and recrystallisation (e.g. from ethanol or aqueous ethanol for DNPs) must be carried out. In that way sharp melting temperatures close to the literature values can be obtained whereas impure samples show depressed melting temperatures and melt over a significant range. Finally students should be aware of the necessary safety precautions involved. In particular they should be conscious of the toxicity and flammability of the compounds concerned and of any corrosive or carcinogenic properties.

(f) Students should be able to propose a sequence of reactions, not exceeding three in number, drawn from the syllabus, to bring about specified conversions of organic compounds. They should be able also to give the essential reaction conditions for such transformations, e.g. solvent, temperature, and to note any other products obtained, e.g. gases evolved. (Examples involving sequences covered by this learning outcome will be found amongst the past papers and model answers mentioned above.)

(g) Students should be able to deduce theoretical preparative yields for stipulated reactions from the syllabus. For this purpose they should be able to recall and write down the stoichiometry of the required process and to realise that the calculation of a theoretical yield must be carried out on a mole basis i.e. the determination of the relationship between the number of moles of starting compound and the number of moles of product. Students are especially urged to note that such calculations cannot and must not be attempted simply on a mass only basis (i.e. by dividing the mass of product by the mass of starting material) and a mole basis is essential. (See for example A2, Q7 (b)(ii), 1995, in which many students incorrectly wrote yield = mass of product × 100 / mass of starting material(s), and the model answer to the same, in which the proper approach is described.) Students should then also be able to calculate actual yields on the basis of that theoretically attainable.

(h) This learning outcome is identical to outcome (l) of Topic 12(q.v.) but is repeated here because of its obvious relevance to the analysis and characterisation of organic compounds.

(i),(j) Students should be able to recall the outline chemistry of the two processes described in the syllabus. Students should appreciate that the first of these is an example of a condensation polymerisation (compare and contrast with addition polymerisation, Topic 11(e)) in which condensation a molecule of water is eliminated between the carboxylic acid and hydroxy functions. Here both reactants, benzene-1,4-dicarboxylic acid and ethane-1,2-diol, have an active function at each end of the molecule so that esterification between the −COOH and −OH functions may take place indefinitely to form long chain polymeric units. This is an example of polyesterification and the product in this case is the polyester known as 'terylene': in this case the trade name is derived from the trivial name of benzene-1,4-dicarboxylic acid which was formerly known as terephthalic acid. Students should have some awareness of the properties of such polyester materials and their relationship to their structures.

The other synthesis included under this outcome is the preparation of a typical pharmaceutical product, in this case that of aspirin (2-ethanoyloxybenzenecarboxylic acid). Students should understand each step of this synthesis, as shown diagramatically below.

$$\text{PhOH} \xrightarrow{\text{NaOH}} \text{PhONa} \xrightarrow{CO_2, \text{ then } H^+} \text{2-HO-C}_6\text{H}_4\text{-COOH}$$

$$\text{2-CH}_3\text{COO-C}_6\text{H}_4\text{-COOH} \xleftarrow{(CH_3CO)_2O}$$

Students should appreciate that the insertion of the carboxylic acid group adjacent to the −OH function in step 2 is a highly specific reaction which is not to be assumed to be of general applicability and should also (c.f. Topic 15.3 (b)(iii) and (c)) understand the nature of the ethanoylation in step 3.